AF618253

A Variational Approach to Fracture and Other Inelastic Phenomena

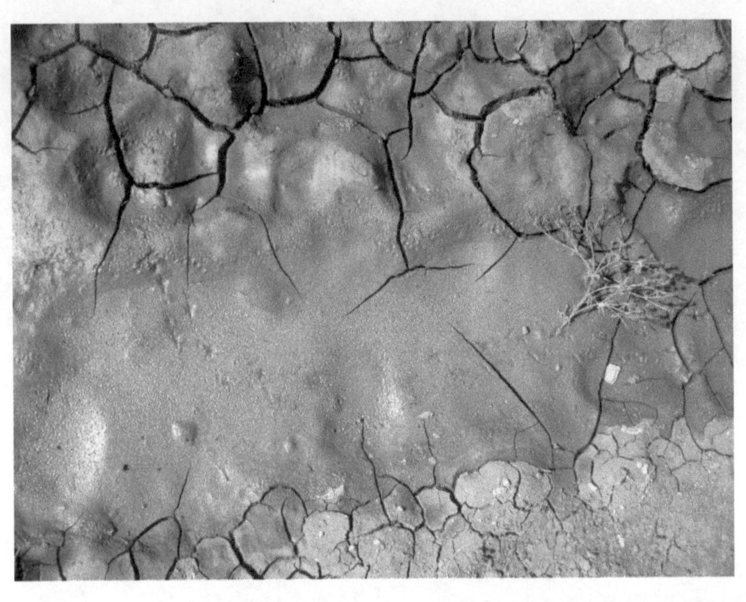

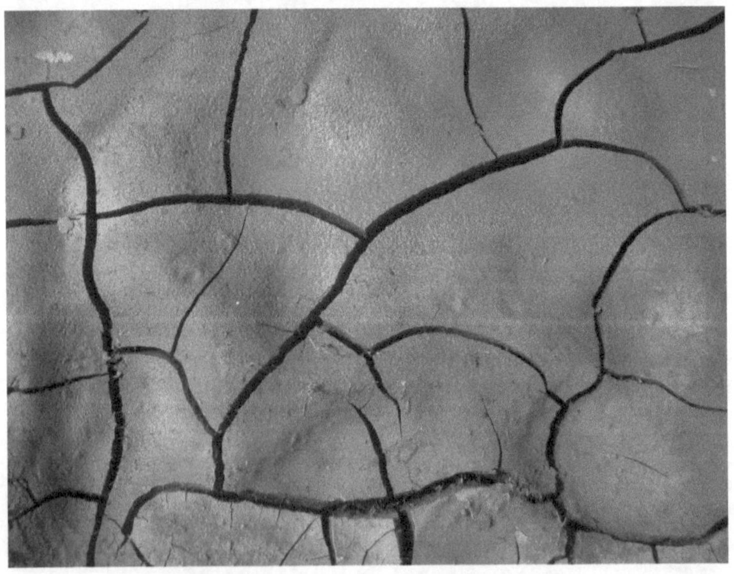

Gianpietro Del Piero

A Variational Approach to Fracture and Other Inelastic Phenomena

Foreword by Roger Fosdick

Previously published in *Journal of Elasticity* Volume 112, Issue 1, 2013

Springer

Gianpietro Del Piero
Dipartimento di Ingegneria
Università di Ferrara
Ferrara, Italy

ISBN 978-94-024-0736-5
Springer Dordrecht Heidelberg New York London

Softcover reprint of the hardcover 1st edition 2014

Printed on acid-free paper

Springer is part of Springer Science+Business Media (www.springer.com)

Dedicated to the memory of my grand-father
Giovanni Fantella (Ivo Fantela)
Lagosta (Lastovo), Dalmatia, 1880–1953

Contents

Foreword

Roger Fosdick

One of the goals of the *Journal of Elasticity: The Physical and Mathematical Science of Solids* is to identify and to bring to the attention of the physical and mathematical sciences research community masterful expositions which contain creative ideas, new approaches and current developments in modeling the behavior of materials. The phenomenon of fracture has enjoyed the attention of fundamental and applied researchers for many years. Its influence on, and responsibility in part for, major growth in fundamental aspects of solid mechanics from the physical modeling of the phenomenon itself, to the development of mathematical techniques for handling singularities, to the establishment of computational methods at large and small scales, and to the development of novel experimental strategies and techniques, is well-known. In relatively recent years, a common theoretical basis and unification has begun to emerge in the fields of fracture, plasticity and damage; a focus on the phenomenon of stability has driven this emergence. The work presented here shows several fundamental and important advances toward this aim, and it presents proposals on how this may be accomplished.

This invited article by Gianpietro Del Piero develops in a masterful way the subject of quasi-static brittle and diffuse fracture from a unified variational point of view based on 'incremental energy minimization'. The article begins with the classical theory of brittle facture due to Griffith. The theory is then regularized so as to include dissipation and elastic unloading via the introduction of surfacial cohesive energies due to Barenblatt and Dugdale. An intriguing description of the formation of microstructure is given and the concept of irreversibility is examined. A diffusive 'process zone' then is introduced into the theory. This generalization recognizes that the fracture energy accompanied with crack growth is not always surfacial in nature, but also can include the particles of the body away from the crack. The assumed existence of a process zone establishes a strong connection with the theories of plasticity and damage. With this, it is then made evident that the incompatibility between the classical theory of plasticity based on Drucker's postulate and the phenomenon of strain softening can be completely resolved with the introduction of a non-local cohesive energy term of the gradient type.

The presentation is comprehensive, written in the style of a series of lectures, one building upon the next. The article includes a large bibliography; at the end of each lecture helpful comments and pertinent references and interpretations regarding modern developments are offered. In order to enlarge the unified theory to the greatest extent and to avoid complex mathematical analysis, the treatment is restricted to one-dimension. The main message is that fracture, plasticity, damage and the creation of microstructure have common theoretical bases, and that they can be treated in a unified way using 'incremental energy minimization' as a basic analytical tool. The non-trivial challenge for the future is to consider this possibility in a two or three dimensional setting.

J Elast (2013) 112:1–2
DOI 10.1007/s10659-013-9443-4

Preface

Gianpietro Del Piero

Published online: 9 May 2013

The present Notes are a re-elaboration of the lectures delivered at the First Sperlonga Summer School on Mechanics and Engineering Sciences, organized by the *Fondazione Tullio Levi-Civita* of Cisterna di Latina, and held in Sperlonga, not far from Rome, in September 2011. In the year and a half elapsed after the Course, the material has been revised and enriched with new and partially unpublished results. Significant additions have been introduced in the occasion of the course *The variational approach to fracture and other inelastic phenomena*, delivered at SISSA, Trieste, in March 2013.

The Notes reflect a research line carried on by the writer over the years, addressed to a comprehensive description of the many aspects of the phenomenon of fracture, and to its relations with other phenomena, such as the formation of microstructure and the changes in the material's strength induced by plasticity and damage. Roughly, the first four Lectures summarize the contents of the paper [4], and Lectures 5 and 6 reflect the subject treated in the paper [5] and successive developments.

A unifying element of these Notes is the concept of cohesive energy. This idea is already explicit in Galilei's book [7] who, in 1638, gave birth to the new science of the strength and rupture of solids. In the following century, this concept was dominant in the mathematical theories of the structure of matter based on interatomic interactions [2], and in Coulomb's theory of rupture [3]. In the second half of the 20th century, the cohesive energy of Barenblatt [1] and Dugdale [6] provided a fruitful regularization of the pre-existing theory of Griffith [8] for brittle fracture.

After a presentation of Griffith's theory in Lecture 1, in Lecture 2 the Barenblatt-Dugdale regularization is discussed. In Lecture 3 it is shown that the concept of cohesive energy can be successfully used to describe the formation of microstructure. A description of the phenomenon of elastic unloading, based on the assumption of dissipativity of the cohesive energy, is given in Lecture 4. A sharp change comes in Lecture 5, in which the cohesive energy, instead of being defined on surfaces as usual, is supposed to be diffused over the

G. Del Piero (✉)
Dipartimento di Ingegneria, Università di Ferrara, 44100 Ferrara, Italy
e-mail: dlpgpt@unife.it

volume. With this assumption, a description of the process zones which usually accompany the crack growth is included into the model. The result is almost identical to the classical plasticity theory based on Drucker's material stability postulate. In Lecture 6 a description of the strain-softening phenomenon, not included in the preceding model, is obtained by adding to the cohesive energy a non-local term of the gradient type.

The whole analysis is carried on using a single mathematical tool, the incremental energy minimization. This explains the uniform structure of all Lectures: the definitions of equilibrium and stability based on the signs of the first and second variation of the energy, followed by the study of the quasi-static evolution and by the analysis of the response curves. In each Lecture, reference and comments are collected in a final Section.

The bibliography is far from complete, and the writer apologizes for unintentional omissions. The criterion followed for citations was to privilege two types of papers: those marking the origin of an idea or of a solution technique, and those providing a, possibly recent, overview on a specific subject.

An evident limit for these Notes is that the whole treatment is one-dimensional. This choice has been made to enlarge the field of the theory as much as possible, keeping away from the complex mathematical apparatus involving, for instance, non-regular fracture sets, non-convex elastic energies, and technically complicated function spaces. An extension to higher dimension is far from trivial. It could be a very challenging perspective for future work.

Udine, Italy, April 2013

References

1. Barenblatt, G.I.: The formation of equilibrium cracks during brittle fracture. General ideas and hypotheses, axially-symmetric cracks. J. Appl. Math. Mech. **23**, 622–636 (1959)
2. Boscovich, R.G.: Theoria Philosophiæ Naturalis: Redacta ad Unicam Legem Virium in Natura Existentium. Vienna (1758), Venice (1763), Paris (1765)
3. Coulomb, C.A.: Essai sur une application des règles *de Maximis & Minimis* à quelques Problèmes de Statique, relatifs à l'Architecture. In: Mémoires de Mathématique & de Physique, présentés à l'Académie Royale des Sciences par divers Savans, vol. 7, pp. 343–382 (1773). Paris (1776)
4. Del Piero, G., Truskinovsky, L.: Elastic bars with cohesive energy. Contin. Mech. Thermodyn. **21**, 141–171 (2009)
5. Del Piero, G., Lancioni, G., March, R.: A diffuse energy approach for fracture and plasticity: the one-dimensional case. J. Mech. Mater. Struct. (2013)
6. Dugdale, D.S.: Yielding of steel sheets containing slits. J. Mech. Phys. Solids **8**, 100–104 (1960)
7. Galilei, G.: Discorsi e Dimostrazioni Matematiche Intorno à Due Nuove Scienze. Elsevier, Leyden (1638)
8. Griffith, A.A.: The phenomenon of rupture and flow in solids. Philos. Trans. R. Soc. Lond. A **221**, 163–198 (1921)

J Elast (2013) 112:3–77
DOI 10.1007/s10659-013-9444-3

A Variational Approach to Fracture and Other Inelastic Phenomena

Gianpietro Del Piero

Received: 11 August 2012 / Published online: 9 May 2013

Abstract The aim of the present article is a description of the many phenomenological aspects of fracture and of their relations with other inelastic phenomena, such as the formation of microstructure and the changes in the material's strength induced by plasticity and damage.

The whole analysis is carried on using a single mathematical tool, the incremental energy minimization. The energy functional is assumed to be made of two parts, an elastic energy and a cohesive energy. The structure assumed for the cohesive energy term determines different modes of inelastic response and fracture. To avoid the heavy technical difficulties met in higher dimension, the whole analysis is one-dimensional.

This article reflects the contents of six lectures delivered in a couple of occasions. The object of the first lecture is Griffith's theory of brittle fracture. In the second lecture, the Barenblatt–Dugdale regularization is discussed. In Lecture 3 it is shown that the concept of cohesive energy can be successfully used to describe the formation of microstructure. A description of the phenomenon of elastic unloading, based on the assumption of dissipativity of the cohesive energy, is given in Lecture 4. The last two lectures deal with the diffuse cohesive energy model. In it, the cohesive energy, instead of being defined on surfaces as usual, is supposed to be diffused over the volume. In particular, the non-local model discussed in Lecture 6 provides a comprehensive description of the strain-softening phenomenon.

Keywords Fracture · Cohesive energies · Variational fracture · Quasi-static evolution · Incremental energy minimization

Mathematics Subject Classification 74G65 · 74R10 · 74N15 · 49J40 · 49K15

G. Del Piero (✉)
Dipartimento di Ingegneria, Università di Ferrara, Ferrara, Italy
e-mail: dlpgpt@unife.it

G. Del Piero
International Research Center M&MOCS, Cisterna di Latina, Italy

1 Lecture 1. Brittle Fracture

1.1 Two Elementary Problems in the Calculus of Variations

To introduce the subject, we start from an elementary problem of one-dimensional elasticity. This gives us the opportunity of fixing the basic notation and terminology.

Consider a one-dimensional elastic bar of length l, subject to axial displacements u. With each displacement associate the elastic strain energy

$$E(u) = \int_0^l w\big(u'(x)\big)dx, \tag{1.1}$$

where $u'(x)$ is the axial strain at the point x, and w is the energy density per unit length of the bar. Assume that w is C^2, strictly convex, and that

$$w(0) = w'(0) = 0. \tag{1.2}$$

Then $w(u')$ is positive for all $u' \neq 0$, w' is strictly increasing, and $w''(u')$ is positive for all u'.

If, as we assume, there are no applied loads, E is the total energy of the bar. For the domain of definition of E we take the set of all continuous functions $u : (0, l) \to \mathbb{R}$ with a summable derivative u'. The elements of this set will be called the *configurations* of the bar.

At the endpoints of the bar, the axial displacements

$$u(0) = 0, \qquad u(l) = \beta l \tag{1.3}$$

are prescribed. Here βl is the total elongation, and β is the relative elongation. The condition $u(l) = \beta l$ will often be written in the integral form

$$\int_0^l u'(x)dx = \beta l. \tag{1.4}$$

In the following, β will be called the *load*, and the configurations which satisfy conditions (1.3) for a given β will be called the configurations *corresponding to* β.

Let now β be fixed. A configuration u corresponding to β is a *global energy minimizer* if $E(v) \geq E(u)$ for all configurations v corresponding to the same β. By setting $v = u + \eta$, we see that u is a global minimizer if the difference

$$E(u+\eta) - E(u) = \int_0^l \Big(w\big(u'(x) + \eta'(x)\big) - w\big(u'(x)\big)\Big)dx \tag{1.5}$$

is non-negative for all *admissible perturbations*, that is, for all functions η with the same regularity of u, which do not change the total length of the bar

$$\int_0^l \eta'(x)dx = 0. \tag{1.6}$$

By introducing a smallness parameter ε, from (1.5) we deduce the expansion

$$E(u+\varepsilon\eta) - E(u) = \varepsilon \int_0^l w'\big(u'(x)\big)\eta'(x)dx + o(\varepsilon). \tag{1.7}$$

The equilibrium configurations of the bar are identified with the configurations u at which the *first variation* of E

$$\delta E(u,\eta) = \lim_{\varepsilon\to 0}\frac{1}{\varepsilon}\big(E(u+\varepsilon\eta) - E(u)\big) \tag{1.8}$$

is non-negative for all perturbations η. By (1.7), the non-negativeness of the first variation is a necessary condition for an energy minimum at u. Therefore, *all energy minimizers are equilibrium configurations.*

From the two preceding equations it follows that an equilibrium configuration is a configuration u which satisfies the inequality

$$\int_0^l w'\big(u'(x)\big)\eta'(x)dx \geq 0 \tag{1.9}$$

for all admissible perturbations. Because η is admissible if and only if $-\eta$ is admissible, the inequality is in fact an equality. A classical argument in the Calculus of Variations states that this condition is verified only if $w'(u'(x))$ is constant over the bar

$$\sigma = w'\big(u'(x)\big). \tag{1.10}$$

This differential equation is the *Euler equation* associated with the minimum problem, and the constant σ is the *axial force* in the bar. Due to the hypothesis of w' strictly increasing, $w'(u'(x))$ constant implies that $u'(x)$ is constant, that is, that u is a *homogeneous configuration*. Moreover, by (1.4), u' constant and u corresponding to β implies $u' = \beta$. Then, the homogeneous configuration u_β defined by

$$u_\beta(x) = \beta x, \tag{1.11}$$

is the only equilibrium configuration corresponding to a given β.

This configuration is a global energy minimizer. Indeed, by the convexity of w, for all configurations u one has

$$w\big(u'(x)\big) - w(\beta) \geq w'(\beta)\big(u'(x) - \beta\big),$$

so that

$$E(u) = \int_0^l w\big(u'(x)\big)dx \geq lw(\beta) + w'(\beta)\int_0^l \big(u'(x) - \beta\big)dx. \tag{1.12}$$

If u is a configuration corresponding to β, the last integral is zero by (1.4). Therefore,

$$E(u) \geq lw(\beta) = E(u_\beta),$$

that is, u_β is a global minimizer. If w is strictly convex, the above inequality is strict, and u_β is the unique global minimizer. The inequality

$$\int_0^l w\big(u'(x)\big)dx \geq lw\bigg(\frac{1}{l}\int_0^l u'(x)dx\bigg), \tag{1.13}$$

which follows from (1.4) and (1.12), is *Jensen's inequality*. It holds for every convex function w.

Now consider a second problem, in which the same energy E is minimized in a different set. Keeping condition $(1.3)_1$, replace condition $(1.3)_2$ by the unilateral constraint

$$u'(x) \geq \nu(x) \quad \forall x \in (0, l), \tag{1.14}$$

with ν a given function. Recalling that, by assumption, the axial force $w'(u')$ is an increasing function of u', this inequality can be interpreted as a constitutive constraint on the force. For example, $\nu(x) = 0$ corresponds to a bar which does not resist compression, that is, to a string. We wish to find the configuration of minimum energy in the presence of this constraint. Of course, the solution $u'(x)$ is not expected to be a constant. As a consequence, a variable internal force $\sigma(x) = w'(u'(x))$, obtained by applying a suitable distribution of external loads, is expected to be necessary to keep the bar in equilibrium in its configuration of minimum energy.

For this problem, an equilibrium configuration is again characterized by inequality (1.9), with the admissible perturbations η defined by

$$\eta(0) = 0, \qquad u'(x) + \eta'(x) \geq \nu(x) \quad \forall x \in (0, l). \tag{1.15}$$

A necessary condition for equilibrium is that

$$w'\big(u'(x)\big) \geq 0 \tag{1.16}$$

almost everywhere in $(0, l)$. Indeed, if $w'(u'(x))$ were negative on an interval, inequality (1.9) would be contradicted by any admissible perturbation with $\eta'(x)$ positive in that interval and zero outside.

Moreover, $w'(u'(x))$ must be zero at all intervals $\mathcal{I}$ at which inequality (1.14) is strict. Indeed, for every such interval there are admissible perturbations η for which η' is negative in $\mathcal{I}$ and zero outside. For them, inequality (1.9) is satisfied only if $w'(u'(x)) = 0$ in $\mathcal{I}$. As a consequence, either (1.14) is satisfied as an equality, or $w'(u'(x))$ is zero. This alternative is expressed by the *complementarity condition*

$$\big(u'(x) - \nu(x)\big)w'\big(u'(x)\big) = 0. \tag{1.17}$$

The conditions (1.14), (1.16), (1.17) are the *Kuhn-Tucker conditions* associated with the minimum problem. They replace the Euler equation in minimization problems constrained by inequalities.

In our first example, the global minimizer was determined by the Euler equation (1.10), using the assumption that w is strictly increasing. Here, it will be determined using the Kuhn-Tucker conditions plus the assumption that $w'(u')$ has the same sign of u'. Indeed, with this assumption, conditions (1.16) and (1.17) reduce to

$$u'(x) \geq 0, \qquad \big(u'(x) - \nu(x)\big)u'(x) = 0.$$

If $\nu(x) < 0$ only the option $u'(x) = 0$ is possible, because $u'(x) = \nu(x) < 0$ would contradict the first inequality. If $\nu(x) \geq 0$, it must be $u'(x) = \nu(x)$ by condition (1.14). Therefore, in an equilibrium configuration u_ν it must be

$$u_\nu'(x) = \nu^+(x), \qquad \nu^+(x) = \max\{\nu(x), 0\}. \tag{1.18}$$

The function u_ν determined by this relation plus the boundary condition $u_\nu(0) = 0$ is a global minimizer for E. Indeed, for any other configuration u obeying the constraint (1.14)

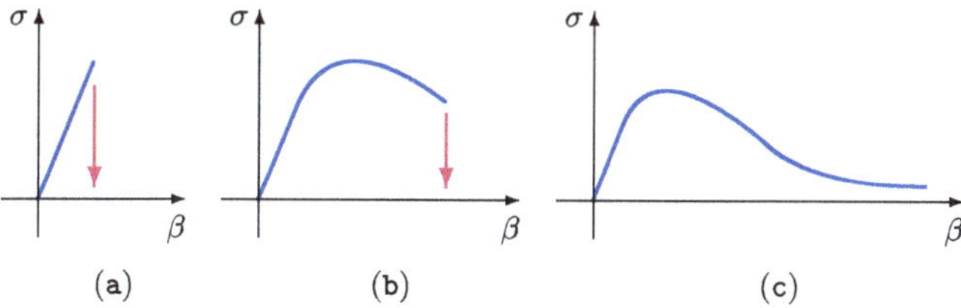

Fig. 1 Fracture modes: brittle (**a**), ductile-brittle (**b**), ductile (**c**)

we have

$$E(u) - E(u_\nu) = \int_0^l \big(w\big(u'(x)\big) - w\big(\nu^+(x)\big)\big)dx.$$

At points at which $\nu(x) \leq 0$ we have $w(\nu^+(x)) = w(0) = 0$, and the right-hand side is positive by the positiveness of w. At points at which $\nu(x) > 0$, from (1.14) we have $u'(x) \geq \nu(x) = \nu^+(x) > 0$, and this implies $w(u'(x)) \geq w(u'_\nu(x))$, because $w(u')$ is increasing for $u' > 0$. Therefore, u_ν is a global minimizer for E.

In the following Sections, appropriate generalizations of the variational procedures described with the two elementary examples given above will be introduced, to formulate some fracture problems as problems of energy minimization.

1.2 A Model for Brittle Fracture

Roughly, two fracture modes are observed in the experiments. Some bodies deform elastically until, without any premonitory sign, a sharp fracture surface appears. When this occurs, the force which can be sustained by the bar suddenly drops to zero. In other cases, a progressive weakening of the material is observed. That is, the force transmitted across the surface gradually reduces to zero, without any catastrophic event. The two fracture modes are called *brittle* and *ductile*, respectively. In most cases, the materials exhibit the intermediate *ductile-brittle* behavior shown in Fig. 1(b), in which the final catastrophic rupture is preceded by an initial regime of progressive weakening.

The model for brittle fracture developed in this Lecture is based on the concepts of fracture point and activation energy. A *fracture point* is a point of the bar at which physical continuity is lost, and the *activation energy* is the energy required to create a fracture point. In the absence of applied loads, the total energy is the sum of the elastic strain energy due to the deformation u' of the unfractured part of the bar, plus the activation energies of the fracture points x_i. The elastic energy has the form (1.1), and the fracture energy is the activation energy θ_a multiplied by the number $\#\{x_i\}$ of the fracture points. Therefore, the total energy is

$$E\big(u, \{x_i\}\big) = \int_0^l w\big(u'(x)\big)dx + \#\{x_i\}\theta_a. \tag{1.19}$$

The domain of E now includes discontinuous functions. We assume that the number $\#\{x_i\}$ is finite, and that in each interval (x_i, x_{i+1}) the function u is continuous and with a continuous derivative. Moreover, at each fracture point the right limit $u(x_i+)$ and the left limit $u(x_i-)$ are assumed to exist. The difference

$$[\![u]\!](x_i) = u(x_i+) - u(x_i-) \tag{1.20}$$

is the *jump* of u at x_i. The jumps are subject to the restriction

$$[\![u]\!](x_i) \geq 0, \tag{1.21}$$

which prevents the interpenetration of the two parts of the bar located at the two sides of a fracture point.

We explicitly assume that, at a fracture point, the jump can be zero. That is, u need not be discontinuous at a fracture point. As a consequence, a function u no longer sufficiently describes a configuration of the bar. Indeed, the same function u may correspond to different energies, depending on the number of fracture points with zero jump amplitude. Due to this circumstance, configurations must be described by pairs $(u, \{x_i\})$, consisting of a displacement function and of a *fracture set*. A configuration is said to be *unfractured* if the fracture set is empty, and *fractured* otherwise.

The boundary condition (1.4) is now replaced by the equation

$$\int_0^l u'(x)dx + \sum_{i=1}^{\#\{x_i\}} [\![u]\!](x_i) = \beta l, \tag{1.22}$$

which tells us that the total elongation βl is the sum of two parts, one due to the bulk deformation u' and one due to the jumps. For a given load β, the functions u *corresponding to* β are the functions with the regularity assumed above, which satisfy Eq. (1.22).

A *perturbation* of a configuration $(u, \{x_i\})$ is a pair $(\eta, \{\xi_j\})$, where η is a function with the same regularity as u, and $\{\xi_j\}$ is a finite subset of $(0, l)$, in general not related to the fracture set $\{x_i\}$. The points ξ_j not in the fracture set are the new fracture points created by the perturbation. By the non-interpenetration condition (1.21) applied to the perturbed configuration $(u + \eta, \{x_i\} \cup \{\xi_j\})$, for all perturbations of $(u, \{x_i\})$ the jump amplitudes $[\![\eta]\!](\xi_j)$ must satisfy the constraint

$$[\![u]\!](\xi_j) + [\![\eta]\!](\xi_j) \geq 0, \tag{1.23}$$

with $[\![u]\!](\xi_j) = 0$ if ξ_j is a newly created fracture point. Note that the jump set $\{x_i\}$ of a configuration is included in the jump set of all perturbed configurations. Therefore, implicit in this definition of perturbation is the idea of the *irreversibility of fracture*, which will be discussed more explicitly later.

1.3 Equilibrium Configurations

An *equilibrium configuration* is a configuration $(u, \{x_i\})$ for which the first variation of the energy

$$\begin{aligned} &\lim_{\varepsilon \to 0} \frac{1}{\varepsilon}\big(E\big(u + \varepsilon\eta, \{x_i\} \cup \{\xi_j\}\big) - E\big(u, \{x_i\}\big)\big) \\ &\quad = \int_0^l w'\big(u'(x)\big)\eta'(x)dx + \theta_a \lim_{\varepsilon \to 0} \frac{1}{\varepsilon}\big(\#\big(\{x_i\} \cup \{\xi_j\}\big) - \#\{x_i\}\big) \end{aligned} \tag{1.24}$$

is non-negative for all *admissible perturbations*, that is, for all pairs $(\eta, \{\xi_j\})$ which satisfy condition (1.23) and leave unaltered the length of the bar

$$\int_0^l \eta'(x)dx + \sum_{j=1}^{\#\{\xi_j\}} [\![\eta]\!](\xi_j) = 0. \tag{1.25}$$

Fig. 2 Energies of the equilibrium configurations $(u, \{x_i\})$ for $\#\{x_i\} = 0, 1, 2, 3$

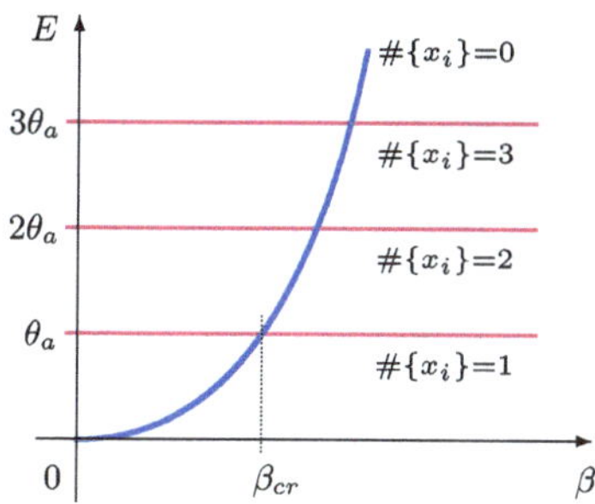

For perturbations with newly created fracture points, the limit on the right-hand side of (1.24) is $+\infty$, and therefore the non-negativeness of the first variation is trivially satisfied. Then, it is sufficient to consider perturbations without newly created fracture points. For such perturbations $\{\xi_j\}$ is a subset of $\{x_i\}$, the limit on the right of (1.24) is zero, and the non negativeness of the first variation is expressed by inequality (1.9).

In particular, for unfractured configurations the fracture set is empty, and therefore only perturbations without jumps need to be considered. Condition (1.25) then reduces to (1.6), and the conclusions of the previous Section for the problem with condition (1.4) hold. That is, $u'(x)$ is constant and equal to β. Therefore, *the only unfractured equilibrium configurations are the pairs* $(u_\beta, \emptyset)$, with u_β the homogeneous deformation (1.11).

For fractured equilibrium configurations, for perturbations without newly created fracture points the non-negativeness of the first variation (1.24) requires

$$0 \le w'(u') \int_0^l \eta'(x)dx = -w'(u') \sum_{j=1}^{\#\{\xi_j\}} [\![\eta]\!](\xi_j),$$

with the inequality due to the constancy of u' and the equality due to condition (1.25). By (1.23), for $\xi_j \in \{x_i\}$ and $[\![u]\!](x_j) > 0$ the jumps $[\![\eta]\!](\xi_j)$ may have any sign. Then the above inequality is satisfied only if $w'(u') = 0$, that is, only if $u' = 0$. Therefore, *in a fractured equilibrium configuration both the axial force* $\sigma = w'(u')$ *and the bulk deformation* u' *are zero.*

In conclusion, there are two types of equilibrium configurations, the homogeneous unfractured configurations $(u_\beta, \emptyset)$, with

$$u'_\beta = \beta, \qquad \sigma = w'(\beta), \qquad E(u_\beta, \emptyset) = lw'(\beta), \tag{1.26}$$

and the fractured configurations $(u, \{x_i\})$ with

$$u' = 0, \qquad \sigma = 0, \qquad E\big(u, \{x_i\}\big) = \theta_a \#\{x_i\}. \tag{1.27}$$

Each equilibrium configuration is equilibrated with the load β given by (1.22). A map of the energies of the equilibrium configurations corresponding to each load β is given in Fig. 2. The equilibrium configurations form a number of curves in the energy-load plane (E, β), one for each number $\#\{x_i\}$ of fracture points.

While a point in the curve $\#\{x_i\} = 0$ represents a single unfractured configuration $(u_\beta, \emptyset)$, each point on the lines $\#\{x_i\} > 0$ represents an infinity of fractured configurations, since there is an infinity of equilibrium configurations with the same energy, which only differ for the positions of the fracture points.

1.4 Stability

A configuration $(v, \{x_j\})$ is said to be *compatible* with a configuration $(u, \{x_i\})$ if there is a perturbation $(\eta, \{\xi_j\})$ such that

$$\big(v, \{x_j\}\big) = \big(u + \eta, \{x_i\} \cup \{\xi_j\}\big). \tag{1.28}$$

Therefore, $(v, \{x_j\})$ is compatible with $(u, \{x_i\})$ if and only if the fracture set $\{x_i\}$ is included in $\{x_j\}$.

We say that a configuration $(u, \{x_i\})$ corresponding to a given β is a *global minimizer* for E if

$$E\big(v, \{x_j\}\big) - E\big(u, \{x_i\}\big) \geq 0 \tag{1.29}$$

for all configurations $(v, \{x_j\})$ compatible with $(u, \{x_i\})$ and corresponding to the same β. We also say that $(u, \{x_i\})$ is a *local minimizer* if there is a $\delta > 0$ such that inequality (1.29) holds for all configurations $(v, \{x_j\})$ compatible with $(u, \{x_i\})$ and corresponding to the same β, whose distance from $(u, \{x_i\})$ is less than δ.

An energy minimizer, whether local or global, is said to be a *stable* equilibrium configuration. Clearly, this definition of stability depends on the choice of a distance function between configurations. However, for whatever choice, it is true that every global minimizer is also a local minimizer, that the non-negativeness of the first variation (1.24) is necessary for a minimum, and, therefore, that every local minimizer is an equilibrium configuration.

For the present model, there is a certain freedom in the choice of the distance function. For consistency with the developments to come in the following Lectures, for the distance between $(v, \{x_j\})$ and $(u, \{x_i\})$ we take the $W^{1,1}$ distance between v and u:

$$\|v - u\| \doteq \sup_{x \in (0,l)} |v(x) - u(x)| + \sup_{x \in (0,l)} |v'(x) - u'(x)|. \tag{1.30}$$

This is in fact only a pseudo-distance, since it makes no difference between no-fracture points and fracture points with null jump amplitude.

For every equilibrium configuration $(u, \{x_i\})$, by the constancy of u' and the strict convexity of w,

$$\begin{aligned} & E\big(u + \eta, \{x_i\} \cup \{\xi_j\}\big) - E\big(u, \{x_i\}\big) \\ & \quad = \int_0^l \big(w\big(u' + \eta'(x)\big) - w(u')\big) dx + \theta_a\big(\#\big(\{x_i\} \cup \{\xi_j\}\big) - \#\{x_i\}\big) \\ & \quad > w'(u') \int_0^l \eta'(x) dx + \theta_a\big(\#\big(\{x_i\} \cup \{\xi_j\}\big) - \#\{x_i\}\big). \end{aligned} \tag{1.31}$$

In particular, for an unfractured equilibrium configuration $(u_\beta, \emptyset)$ the right-hand side reduces to

$$w'(\beta) \int_0^l \eta'(x) dx + \theta_a \#\{\xi_j\}.$$

For perturbations $(\eta, \emptyset)$ without jumps, this sum is zero by (1.25). For perturbations with jumps, by the inequality

$$\left| \int_0^l \eta'(x) dx \right| \leq l \sup_{x \in (0,l)} |\eta'(x)| \leq l \|\eta\|,$$

this sum is positive if $\|\eta\| < \theta_a / lw'(\beta)$. Therefore, *every unfractured equilibrium configuration is a local minimizer*. Moreover, by the positiveness of w, from the equality in (1.31) it follows that

$$E\big(u_\beta + \eta, \{\xi_j\}\big) - E(u_\beta, \emptyset) \geq -lw(\beta) + \theta_a.$$

Then $(u_\beta, \emptyset)$ is a global minimizer if $lw(\beta) \leq \theta_a$, that is, if $\beta \leq \beta_{cr}$, where

$$\beta_{cr} = w^{-1}(\theta_a / l) \tag{1.32}$$

is the critical value of β at which the energies of the equilibrium configurations with $\#(x_i) = 0$ and $\#(x_i) = 1$ are equal, see Fig. 1. Conversely, if $\beta > \beta_{cr}$ the configuration $(u_\beta, \emptyset)$ is not a global minimizer, because the configuration $(v, \{x_j\})$ with $v' = 0$ and $\#\{x_j\} = 1$ corresponding to the same β has a lower energy. Therefore, *$(u_\beta, \emptyset)$ is a global minimizer if and only if $\beta \leq \beta_{cr}$*.

For a fractured equilibrium configuration $w'(u')$ is zero, and the right-hand side of (1.31) is non-negative for all perturbations. Then, *every fractured equilibrium configuration is a global minimizer.*

It may seem contradictory, that for $\beta \leq \beta_{cr}$ there are two global minimizers, one fractured and one unfractured, with different energies. Actually, the two are global minimizers in two different sets, each set being made of all configurations compatible with the given configuration.

This shows that, in fracture, indiscriminate global minimization is meaningless: at least, one has to decide whether to minimize starting from a fractured or from an unfractured configuration. This naturally leads to consider the evolution of the energy minimizers in a bar subject to a varying load.

1.5 Quasi-Static Evolutions

A *load process* is a continuous real-valued function $t \mapsto \beta_t$. The parameter t can, but need not, be identified with the physical time. Indeed, the process is supposed to be sufficiently slow to be rate-independent. That is, the rate effects, such as, for example, inertia and viscous dissipation, can be neglected. In spite of this, β_t cannot be identified with t. Indeed, we wish to be free to increase and decrease the load, that is, to subject the bar to *loading* and *unloading* regimes. Only in the study of a single loading regime the identification of β_t with t is possible.

A *deformation process* corresponding to a given load process $t \mapsto \beta_t$ is a continuous path $t \mapsto (u_t, \{x_{it}\})$ in the configuration space such that, for each t, $(u_t, \{x_{it}\})$ is a configuration corresponding to β_t. A deformation process is required to satisfy the two following conditions:

(i) for all $\tau > t$, $(u_\tau, \{x_{i\tau}\})$ is compatible with $(u_t, \{x_{it}\})$,
(ii) for all t,

$$\dot{E}\big(u_t, \{x_{it}\}\big) \leq \sigma_t l \dot{\beta}_t. \tag{1.33}$$

The first condition is equivalent to

$$\{x_{it}\} \subseteq \{x_{i\tau}\} \quad \forall \tau > t. \tag{1.34}$$

This is a condition of *irreversibility of fracture*: in a deformation process, fracture points can be created, but cannot be eliminated. In condition (ii), the superimposed dots denote differentiation with respect to t, and the right-hand side is the power supplied from the exterior. Inequality (1.33) states that for every $\varepsilon > 0$ the variation of the energy in any interval $(t, t+\varepsilon)$ cannot exceed the power supplied in the same interval. This is a combination, in a purely mechanical context, of the first two laws of thermodynamics: the energy balance and the Clausius-Duhem inequality [77, Eq. (81.7)]. We say that a configuration $(u, \{x_i\})$ is *accessible* from $(u_o, \{x_{io}\})$ if there is a load process $t \mapsto \beta_t$ to which corresponds a deformation process $t \mapsto (u_t, \{x_{it}\})$ from $(u_o, \{x_{io}\})$ to $(u, \{x_i\})$. The compatibility condition (i) and the energetic dissipation inequality (ii) will be called the *accessibility conditions*.

An *equilibrium process* is a deformation process in which every configuration is an equilibrium configuration, and a *quasi-static evolution* is an equilibrium process in which every configuration is a stable equilibrium configuration. Because each configuration $(u_t, \{x_{it}\})$ corresponds to a β_t in the corresponding load process, in a quasi-static evolution each configuration is a local or global minimizer among the configurations corresponding to the same β_t. This definition somehow reflects the idea that, under a varying load and in the absence of rate effects, the bar's deformation evolves following a path of least energy.

In the forthcoming Lectures, a major task will be to determine the quasi-static evolutions, if they exist, corresponding to a given load process. In the next Section we prove the rather surprising fact that, in the model for brittle fracture considered here, there is no quasi-static evolution involving a change of the fracture set.

1.6 Crack Initiation

Consider a quasi-static evolution from the natural configuration $(0, \emptyset)$, corresponding to a monotonically increasing load process. For β less than the critical value β_{cr} given by (1.32), the evolution follows the homogeneous deformations $(u_\beta, \emptyset)$, each of which is the global energy minimizer for the corresponding β. A delicate point is to decide what happens at $\beta = \beta_{cr}$. Indeed, there are two alternatives:

- a fracture point is created, and the evolution follows the global minima for $\#\{x_i\} = 1$,
- no fracture point is created, and the evolution follows the local minima for $\#\{x_i\} = 0$.

The creation of a fracture point requires a transition from the homogeneous configuration $(u_{\beta_{cr}}, \emptyset)$ to a fractured equilibrium configuration $(u_1, \{x_1\})$, with x_1 any point in $(0, l)$, with u_1 the piecewise constant function

$$u_1(x) = \begin{cases} 0 & \text{if } x < x_1, \\ \beta_{cr} l & \text{if } x > x_1, \end{cases}$$

and with β keeping the constant value β_{cr}. This is necessarily a non-equilibrium transition. Indeed, there are no unfractured equilibrium configurations corresponding to β_{cr} besides $(u_{\beta_{cr}}, \emptyset)$, and for whatever x_1 the fractured equilibrium configuration $(u_1, \{x_1\})$ corresponding to β_{cr} is far away from $(u_{\beta_{cr}}, \emptyset)$, because $u_1' = 0$ and $u_{\beta_{cr}}' = \beta_{cr}$ implies

$$\|u_1 - u_{\beta_{cr}}\| \geq \sup_{x \in (0,l)} |u_1'(x) - u_{\beta_{cr}}'(x)| = \beta_{cr}.$$

In all non-equilibrium transitions from $(u_{\beta_{cr}}, \emptyset)$ to $(u_1, \{x_1\})$, inequality (1.33) is violated. Indeed, at each instant t the power supplied from the exterior is zero because the transition occurs at constant β, while the energy E has a positive jump of amount θ_a at the instant at which the crack is formed.

This leads to the conclusion that in the present model *the fractured configurations are inaccessible from the unfractured configurations.* This conclusion is made more explicit by saying that the jump of the fracture energy at the crack initiation creates an *energy barrier* of height θ_a, which forbids the continuation along the curve of the global minima. Note that the converse also is true: by the condition (1.34) of irreversibility of fracture, the unfractured configurations are inaccessible from the fractured configurations.

1.7 Comments

This first Lecture is essentially a one-dimensional version of the variational approach established in [69] by Francfort and Marigo for the Griffith model of fracture. In one dimension, much of the problem is trivialized: since the cracks concentrate on isolated points, there is no crack length and no crack growth. And the fact that as soon as a crack is created the force drops to zero and remains equal to zero forever renders uninteresting the study of the fractured regime. And yet, the one-dimensional problem is rich enough to deserve analysis. Mainly, because it helps understanding the basic structure of the fracture problems, keeping away from the technical difficulties encountered in higher dimension.

Let us recall, and briefly comment, the assumptions made and the main concepts introduced in this Lecture.

1. The assumption of *convexity* of the strain energy density w has the purpose of avoiding phenomena of phase transition within the elastic regime. Though the study of fracture in the presence of phase transition is an interesting subject, it will not be treated in these Notes.

2. A minimization restricted to compatible configurations is required by the *irreversibility of fracture* [69]. This is the property that, in any deformation process, the fracture set at time t must be included in the fracture set at all times $\tau > t$. As a consequence, a fracture cannot heal. A minimization under this constraint is called *unilateral* by some authors [24, 40, 94, 101].

3. For the *distance* between configurations, different choices have been made, such as the distance generated by the norm of H^1, [47, 94], and of BV, [43, 54, 63]. The $W^{1,1}$ distance (1.30) was introduced in [57]. As we shall see in Lecture 3, this seems to be the most appropriate choice in view of a more general theory including the presence of microsructure.

4. As shown in the preceding Section, in Griffith's model it is impossible to satisfy the energetic dissipation inequality (1.33) at the creation of a new fracture point. This point is crucial, since it prevents the possibility of describing crack initiation with Griffith's model. In the literature, there is a general consensus on this point [41, 69, 94, 97]. The supporting arguments generally refer to a two- or three-dimensional setting. The argument used in the previous Section, based on the concept of accessibility, is appropriate to the present one-dimensional context.

In Francfort and Marigo's original model [69], the energy inequality (1.33) is absent and the only requirement for accessibility is the irreversibility of fracture. In this context, it is possible to include crack opening in a one-dimensional model. Some arguments in favor of the use of global minimization in the description of crack opening in higher dimension have recently been raised in [101].

5. A way for relaxing the energetic condition (1.33) is provided by the notion of *ε-accessibility* introduced in [94]. For a fixed $\varepsilon > 0$, a configuration $(u_1, \{x_1\})$ is said to be ε-accessible from $(u_0, \{x_0\})$ if the two configurations can be joined by a continuous path in the configuration space, such that

$$E\big(u_\tau, \{x_\tau\}\big) - E\big(u_t, \{x_t\}\big) \leq \varepsilon \quad \forall \tau > t. \tag{1.35}$$

This property is important to preserve in the continuous limit the stability properties of approximate solutions evaluated over finite time increments. To agree with the general requirement (1.33), this condition presupposes an extra supply of energy, which makes it possible to overcome energy barriers of height ε. In [94], the nature of this energy is not specified. A possibility is that this energy be released at the microscopic level, due to some change in the internal structure of the material.

6. The inability of Griffith's approach to describe crack initiation is due to the excessive simplicity of the model. A more realistic response can be obtained by assuming less brutal mechanisms of crack opening. As shown in the following Lecture, in the Barenblatt-Dugdale model this is made by taking into account the resistance to crack opening offered by the internal cohesion of the material.

1.8 Historical Note

Due to its catastrophic character, brittle fracture has been the first and the most important object of investigation in fracture mechanics. For a long time, fracture was considered a mysterious and unpredictable phenomenon. Some light eventually came from the idea of attributing an energy to the cracks, and making a balance between the energy spent and the elastic strain energy recovered in a virtual crack growth.

In two- or three-dimensional models, fracture points are replaced by fracture lines and surfaces, respectively. The idea of a fracture energy proportional to the length of the fracture line or to the area of the fracture surface is due to A.A. Griffith. In his fundamental paper [76], using an incremental energy balance, he evaluated the force necessary for crack growth in cracked plates made of different materials and subjected to various types of loading.

In Griffith's paper, and for decades in the subsequent literature, scarce attention was paid to the problem of crack initiation. Real materials are full of cracks and microcracks, and the important technical problem was how to control the growth of the existing cracks, which, as taught by experience, may suddenly grow, with catastrophic consequences. The merit of revisiting the subject with new eyes goes to G. Francfort and J.J. Marigo, whose paper [69] offers a genuinely variational formulation of the problem, and translates into mathematical language the phenomenological requirement of the irreversibility of fracture.

For the mathematical community, the interest in fracture problems originated from the peculiar structure of Griffith's model which, from the viewpoint of the calculus of variations, is a *free discontinuity problem* [5, 26]. At the end of the 1980's, this class of problems received large attention, as a source of applications for the just established Γ-convergence theory [50]. The first extensively studied free discontinuity problem was the image segmentation problem [107]. A numerical solution technique for this problem based on a regularized representation of the discontinuities was proposed by Ambrosio and Tortorelli [6]. With the proposed technique, the problem is reduced to the solution of a discretized problem in ordinary Sobolev spaces. Numerical simulations giving an idea of the computational efficiency of Ambrosio and Tortorelli's technique can be found in [22, 23], and [55].

It is a merit of the paper [69] to have highlighted the peculiarities of the fracture problem, which make it much more than a mere replica of the image segmentation problem. An outline of the variational approach to brittle fracture and a review of the related literature may be found in the article [24]. Some existence results for the variational problem are collected in the review paper [45].

2 Lecture 2. Regularization

2.1 Cohesive Energies

The cohesive energy models are based on the idea that the fracture energy, instead of being released instantaneously at crack initiation as in Griffith's model, is released gradually with the growth of the crack opening. Because a gradual release presumes some cohesion between the separating parts of a crack, it seems appropriate to call *cohesive* the energy spent in this way.

In the following, the cohesive energy θ is assumed to be a monotonic non-decreasing function of the jump amplitude $[\![u]\!]$. By the non-interpenetration condition (1.21), θ is defined on the positive half-line. The limits

$$\theta_a = \lim_{[\![u]\!]\to 0^+} \theta([\![u]\!]), \qquad \theta_r = \lim_{[\![u]\!]\to +\infty} \theta([\![u]\!]),$$

are the *activation energy* and the *rupture energy*. They are the energies required for *crack initiation* and for *total fracture*, respectively.

In fracture mechanics, cohesive energies were introduced independently by Dugdale [65] and Barenblatt [9]. As shown in Fig. 3(a), in Dugdale's model θ is continuous and piecewise linear, while in Barenblatt's model it is a smooth function. In both models the activation energy is zero, the rupture energy is finite, θ is concave, and

$$\theta(0) = 0, \qquad \theta'(0) > 0. \tag{2.1}$$

In Dugdale's model, total fracture is attained at a finite value $[\![u]\!]_{rd}$ of $[\![u]\!]$. In Barenblatt's model, the jump opening $[\![u]\!]_{rb}$ at total fracture may be finite or infinite. In both cases,

$$\theta([\![u]\!]) = \theta_r \qquad \begin{cases} \forall [\![u]\!] \geq [\![u]\!]_{rd} & \text{in Dugdale's model,} \\ \forall [\![u]\!] \geq [\![u]\!]_{rb} & \text{in Barenblatt's model,} \end{cases} \tag{2.2}$$

where $[\![u]\!]_{rb}$ may be infinite. Griffith's energy is a particular cohesive energy of the form

$$\theta([\![u]\!]) = \begin{cases} 0 & \text{if } [\![u]\!] = 0, \\ \theta_r & \text{if } [\![u]\!] > 0, \end{cases} \tag{2.3}$$

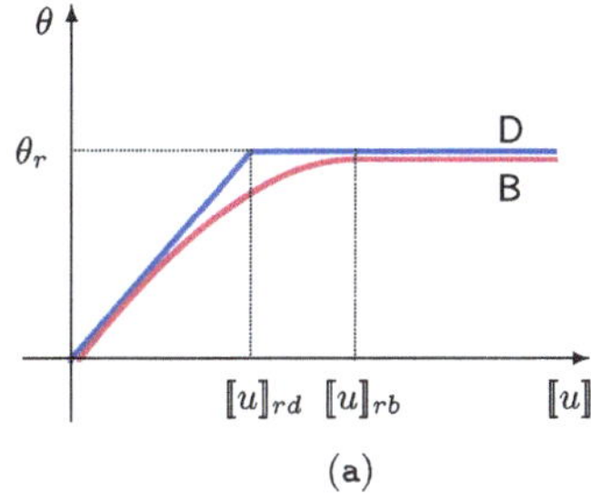

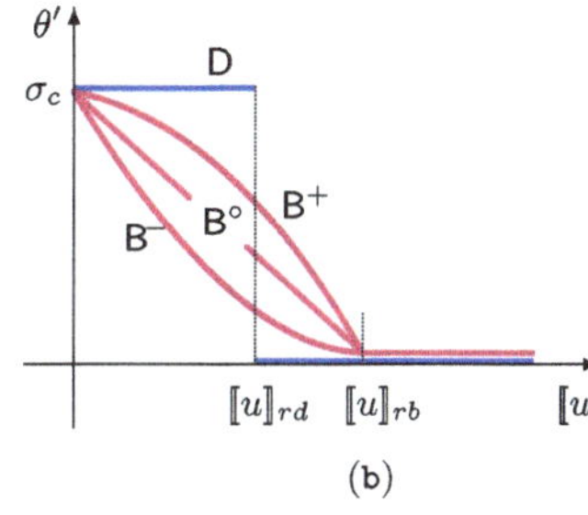

Fig. 3 The cohesive energy θ, (**a**), and the bridging force θ', (**b**), in the models of Dugdale (D) and Barenblatt (B). In (**b**), the curves B^+, B°, B^- correspond to a superquadratic, quadratic, subquadratic energy, respectively

in which $\theta_a = \theta_r$. It can be seen as the discontinuous limit of Dugdale's and Barenblatt's energies for $[\![u]\!]_r \to 0$. In this sense, both energies are regularizations of Griffith's energy.

The derivative $\theta'([\![u]\!])$ is the *bridging force* transmitted across a crack of amplitude $[\![u]\!]$. From Fig. 3(b) we see that in Dugdale's model this force keeps a constant value σ_c up to the critical value $[\![u]\!]_{rd}$ of the crack opening, and then drops to zero. In Barenblatt's model, the dependence of the force on $[\![u]\!]$ is continuous and decreasing. The figure shows that for Barenblatt's energy the force in the interval $(0, [\![u]\!]_{rb})$ can be a concave, linear, or convex function of $[\![u]\!]$. In the three cases we say that the energy is *superquadratic, quadratic, subquadratic*, respectively.

While Griffith's model relies on the concepts of activation energy and fracture point, in the Barenblatt-Dugdale model there is no activation energy, and the concept of a fracture point must be conveniently refined. Indeed, in the cohesive energy model there is a particular situation, here called of *pre-fracture*, in which a crack is open, and yet a non-null bridging force can be transmitted across the crack.

The physical continuity of a bar is supposed to be lost only at total fracture, when the jump amplitude reaches the critical value $[\![u]\!]_r$ and the bridging force drops to zero. When this occurs a *fracture point* in the sense of Griffith's model is created. The fracture becomes irreversible, and in all subsequent deformation processes the cohesive energy keeps the constant value θ_r and the bridging force is zero.

Therefore, in the present one-dimensional setting there are three types of points, *no-fracture, pre-fracture*, and *fracture points*. Since the activation energy is zero, there is no energy cost for a transition from a no-fracture to a pre-fracture point. That is, a pre-fracture point with zero jump amplitude is energetically equivalent to a no-fracture point.

Like in Griffith's model, a configuration is defined by a pair $(u, \{x_i\})$, formed by a displacement function and a fracture set. However, due to the equivalence between no-fracture points and pre-fracture points with jumps of zero amplitude, all pre-fracture points are jump points of u. Therefore, only the fracture points need to be specified in the fracture set.

Thus, there are essentially two types of configurations, the *pre-fractured configurations*, for which the fracture set is empty, and the *totally fractured configurations*, for which the fracture set is non-empty. A pre-fractured configuration is fully described by the displacement function u, while a totally fractured configuration requires the specification of the set of the fracture points. A pre-fractured configuration is *unfractured* if the jump set of u is empty, and is *partially fractured* if the jump set of u is non-empty.

Like in Griffith's model, in the one-dimensional case the study of post-fracture behavior is of scarce interest. In what follows, we concentrate on pre-fractured configurations. The total energy of a pre-fractured configuration is

$$E(u) = \int_0^l w\big(u'(x)\big)dx + \sum_{i=1}^{\#u} \theta\big([\![u]\!](x_i)\big). \tag{2.4}$$

Its domain of definition is the set of all functions $u : (0, l) \to \mathbb{R}$ with a finite, possibly zero, number $\#u$ of jump points x_i, of amplitude $0 < [\![u]\!](x_i) < [\![u]\!]_r$. The regularity is the same required in Griffith's model: u is supposed to be continuous and with a continuous derivative in all intervals (x_i, x_{i+1}), and to have a right and a left limit at each x_i.

2.2 Equilibrium Configurations

For the present model, u is a configuration *corresponding to* β if it satisfies the boundary condition (1.22), and a *perturbation* for u is a function η with the same regularity required as for u. Consider the perturbed configuration $u+\varepsilon\eta$, where ε is a smallness parameter. In the difference

$$E(u+\varepsilon\eta)-E(u)$$
$$=\varepsilon\left(\int_0^l w'\big(u'(x)\big)\eta'(x)dx+\sum_{j=1}^{\#\eta}\theta'\big([\![u]\!](x_j)\big)[\![\eta]\!](x_j)\right)+o(\varepsilon), \tag{2.5}$$

the term between parentheses is the first variation $\delta E(u,\eta)$. By definition, u is an equilibrium configuration if $\delta E(u,\eta)$ is non-negative for all *admissible perturbations*, that is, for all perturbations which satisfy the non-interpenetration condition (1.23) for the given u, and the boundary condition (1.25).

In particular, for a perturbation without jumps the non-negativeness of $\delta E(u,\eta)$ is expressed by inequality (1.9), and condition (1.25) reduces to (1.6). This implies that both $w'(u')$ and u' are constant. As in (1.10), $w'(u')$ is identified with the axial force σ. Then, using (1.25), the first variation at an equilibrium configuration u can be given the simpler form

$$\delta E(u,\eta)=\sum_{j=1}^{\#\eta}\big(\theta'\big([\![u]\!](x_j)\big)-\sigma\big)[\![\eta]\!](x_j). \tag{2.6}$$

Now take a perturbation with a single jump at a point x_j at which $[\![u]\!](x_j)=0$. By (1.23), the jump amplitude $[\![\eta]\!](x_j)$ must be non-negative. Then the first variation is non-negative only if $\theta'([\![u]\!](x_j))=\theta'(0)\geq\sigma$. That is, in an equilibrium configuration the axial force is bounded from above by

$$\sigma\leq\theta'(0). \tag{2.7}$$

Finally, assume that u has a jump at x_i, and take a perturbation with a single jump at x_i. In this case, by (1.23), in an admissible perturbation the sign of $[\![\eta]\!](x_i)$ is not restricted. Then the non-negativeness of $\delta E(u,\eta)$ for all η requires

$$\sigma=\theta'\big([\![u]\!](x_i)\big) \tag{2.8}$$

at every jump point of u. Conversely, if conditions (2.7) and (2.8) are satisfied, the first variation is non-negative.

Thus, in the Barenblatt-Dugdale model a pre-factured equilibrium configuration is a function with constant u', such that $\sigma=w'(u')$ satisfies conditions (2.7) and (2.8). That is, the axial force at the no-fracture points coincides with the bridging force $\theta'([\![u]\!](x_i))$ at each pre-fracture point, and both are bounded from above by $\theta'(0)$.

By (2.8) and by the assumed positiveness of θ', σ is positive in all partially fractured configurations. In view of (2.7), we conclude that in all partially fractured equilibrium configurations the axial force is subject to the two-side bound

$$0<\sigma\leq\theta'(0). \tag{2.9}$$

Moreover, because $\sigma=w'(u')$, w' is strictly increasing, and $w'(0)=0$, the axial strain is bounded by

$$0<u'\leq\beta_c,\qquad \beta_c=(w')^{-1}\big(\theta'(0)\big). \tag{2.10}$$

For an unfractured configuration, by condition (1.22), the constancy of u' implies $u' = \beta$. That is, like in Griffith's model, the only unfractured equilibrium configuration corresponding to β is the homogeneous configuration $u_\beta(x) = \beta x$.

For this configuration, the upper bound (2.7) holds. Then the upper bound in (2.10) holds as well, that is, $\beta = u'_\beta \leq \beta_c$. That is, *$u_\beta$ is an equilibrium pre-fractured configuration only if $\beta \leq \beta_c$*. Moreover, since the homogeneous deformations u_β are the only possible unfractured equilibrium configurations, *all pre-fractured equilibrium configurations corresponding to $\beta > \beta_c$ are partially fractured.*

2.3 Stability

A pre-fractured configuration u corresponding to β is a *global minimizer* if

$$E(v) \geq E(u)$$

for all pre-fractured configurations v corresponding to the same β. It is a *local minimizer* if there is a $\delta > 0$ such that the same inequality holds for all v corresponding to β for which the distance $\|v - u\|$ is less than δ. For the distance $\|\cdot\|$ we keep the definition (1.30). A local minimizer is also called a *stable* equilibrium configuration. Clearly, all global minimizers are also local minimizers.

We begin our stability analysis by proving that all unfractured equilibrium configurations corresponding to $\beta < \beta_c$ are stable.

Proposition 2.1 *Let w be a strictly convex C^2 function obeying conditions* (1.2), *and let θ be any C^2 function with $\theta(0) = 0$ and $\theta'(0) > 0$. Then all unfractured equilibrium configurations u_β with $\beta < \beta_c$ are stable.*

Proof For $u'_\beta = \beta$, by the convexity of w and by (1.25),

$$\int_0^l \big(w\big(\beta + \eta'(x)\big) - w(\beta)\big)dx \geq w'(\beta) \int_0^l \eta'(x)dx = -w'(\beta) \sum_{j=1}^{\#\eta} [\![\eta]\!](x_j).$$

Therefore,

$$\begin{aligned} E(u_\beta + \eta) - E(u_\beta) &= \int_0^l \big(w\big(\beta + \eta'(x)\big) - w(\beta)\big)dx + \sum_{j=1}^{\#\eta} \theta\big([\![\eta]\!](x_j)\big) \\ &\geq \sum_{j=1}^{\#\eta} \big(\theta\big([\![\eta]\!](x_j)\big) - w'(\beta)[\![\eta]\!](x_j)\big), \end{aligned}$$

with all $[\![\eta]\!](x_j)$ positive by (1.23). By the strict monotonicity of w', $\beta < \beta_c$ implies $w'(\beta) < w'(\beta_c) = \theta'(0)$, with the equality due to (2.10). Then there is a $\delta > 0$ such that $\theta'(\eta) > w'(\beta)[\![\eta]\!]$ for all $[\![\eta]\!] < \delta$. Then $E(u_\beta + \eta) \geq E(u_\beta)$ for all perturbations whose largest jump amplitude $[\![\eta]\!](x_j)$ is less than δ. From the definition (1.30) of $\|\cdot\|$ and from (1.25) it follows that

$$\|\eta\| \geq \sup_{x \in (0,l)} |\eta'(x)| \geq \frac{1}{l} \left| \int_0^l \eta'(x)dx \right| = \frac{1}{l} \sum_{j=1}^{\#\eta} [\![\eta]\!](x_j).$$

Then $E(u_\beta + \eta) \geq E(u_\beta)$ for all η with $\|\eta\| < \delta/l$. □

Because u_β is an equilibrium configuration only if $\beta \leq \beta_c$, the only unfractured equilibrium configuration which may not be stable is the configuration corresponding to β_c. For it, stability depends on the shape of θ near the origin, see [63, Sect. 3].

For partially fractured equilibrium configurations, the first variation (2.6) is zero by the equilibrium condition (2.8). Then a necessary condition for stability is the non-negativeness of the second variation

$$\delta^2 E(u,\eta) = w''(u') \int_0^l \eta'^2(x)dx + \sum_{j=1}^{\#\eta} \theta''\big([\![u]\!](x_j)\big)[\![\eta]\!]^2(x_j) \geq 0,$$

for all admissible perturbations η. In particular, for perturbations with constant η' and with a single jump $[\![\eta]\!]$ at a jump point x_i of u, condition (1.25) takes the form $l\eta' + [\![\eta]\!] = 0$, and the second variation reduces to

$$lw''(u')\eta'^2 + \theta''\big([\![u]\!](x_i)\big)[\![\eta]\!]^2 = \big(lw''(u') + l^2\theta''\big([\![u]\!](x_i)\big)\big)\eta'^2.$$

Therefore, a necessary condition for stability is that

$$w''(u') + l\theta''\big([\![u]\!](x_i)\big) \geq 0, \tag{2.11}$$

for all jump points x_i. In general, this condition is not sufficient. In fact, a more restrictive necessary condition holds for strictly concave energies. We recall that a function θ is *concave* if

$$\theta\big(\lambda p + (1-\lambda)q\big) \geq \lambda\theta(p) + (1-\lambda)\theta(q) \quad \forall p,q \geq 0,\ \forall\lambda \in (0,1), \tag{2.12}$$

and that it is *strictly concave* if the above inequality is strict. Both Barenblatt's and Dugdale's energies are concave, but only Barenblatt's energy is strictly concave.

Proposition 2.2 *Let w and θ be as in Proposition* 2.1, *with θ strictly concave in* $(0, [\![u]\!]_r)$. *Then a pre-fractured equilibrium configuration u is stable only if $\#u \leq 1$.*

Proof If θ is strictly concave, θ' is strictly decreasing. Then, by the equilibrium condition (2.8), all jumps have the same amplitude $[\![u]\!]$. Let x_1 and x_2 be jump points for u. Take the perturbation

$$\eta(x) = \begin{cases} 0 & \text{for } 0 < x < x_1, \\ a & \text{for } x_1 < x < x_2, \\ 0 & \text{for } x_2 < x < l, \end{cases}$$

for which

$$E(u+\eta) - E(u) = \theta([\![u]\!]+a) + \theta([\![u]\!]-a) - 2\theta([\![u]\!]) = \theta''([\![u]\!])a^2 + o\big(a^2\big).$$

For θ strictly concave, $\theta''([\![u]\!])$ is negative. Then $E(u+\eta) < E(u)$ for sufficiently small $|a|$. That is, u is not a minimizer for E. □

A sufficient condition for stability is proved below for subadditive cohesive energies. We recall that a function $\theta : (0,+\infty) \to \mathbb{R}$ is *subadditive* if

$$\theta(a) + \theta(b) \geq \theta(a+b) \quad \forall a,b \geq 0. \tag{2.13}$$

A concave function with $\theta(0)=0$ is subadditive. Indeed, from the definition (2.12) of concavity, for

$$p=a+b, \qquad q=0, \qquad \lambda=\frac{a}{a+b},$$

one has

$$\theta(a)\geq\frac{a}{a+b}\,\theta(a+b).$$

Then (2.13) is obtained by interchanging a and b and summing.

Proposition 2.3 *Let w and θ be as in Proposition* 2.1. *If θ is concave, an equilibrium configuration with $\#u=1$ is a local minimizer if*

$$w''(u')+l\theta''([\![u]\!])>0. \tag{2.14}$$

Proof Consider the perturbed energy

$$E(u+\eta)=\int_0^l w\big(u'+\eta'(x)\big)dx+\sum_{j=1}^{\#(u+\eta)}\theta\big([\![u]\!](x_j)+[\![\eta]\!](x_j)\big).$$

By Jensen's inequality (1.13) and by the boundary condition (1.25),

$$\begin{aligned}\int_0^l w\big(u'+\eta'(x)\big)dx &\geq l w\left(u'+\frac{1}{l}\int_0^l\eta'(x)dx\right)\\ &=l w\left(u'-\frac{1}{l}[\![\eta]\!]_{tot}\right)\\ &=l w(u')-w'(u')[\![\eta]\!]_{tot}+\frac{1}{2l}w''(u')[\![\eta]\!]_{tot}^2+o\big([\![\eta]\!]_{tot}^2\big),\end{aligned}$$

where $[\![\eta]\!]_{tot}$ is the sum of all jumps of η, and $[\![u]\!]$ is the unique jump of u. Moreover, for $\#u=1$, by the subadditivity of θ,

$$\begin{aligned}\sum_{j=1}^{\#(u+\eta)}\theta\big([\![u]\!](x_j)+[\![\eta]\!](x_j)\big)&\geq\theta([\![u]\!]+[\![\eta]\!]_{tot})\\ &=\theta([\![u]\!])+\theta'([\![u]\!])[\![\eta]\!]_{tot}+\frac{1}{2}\theta''([\![u]\!])[\![\eta]\!]_{tot}^2+o\big([\![\eta]\!]_{tot}^2\big).\end{aligned}$$

Summing, and recalling that $w'(u')=\theta'([\![u]\!])$,

$$E(u+\eta)\geq E(u)+\frac{1}{2l}\big(w''(u')+l\theta''([\![u]\!])\big)[\![\eta]\!]_{tot}^2+o\big([\![\eta]\!]_{tot}^2\big).$$

Then, the inequality $E(u+\eta)\geq E(u)$ holds for all perturbations η with sufficiently small $|[\![\eta]\!]_{tot}|$, and because $|[\![\eta]\!]_{tot}|=|\int\eta'|\leq l\|\eta\|$, we conclude that $E(u+\eta)\geq E(u)$ for all η with sufficiently small $\|\eta\|$. □

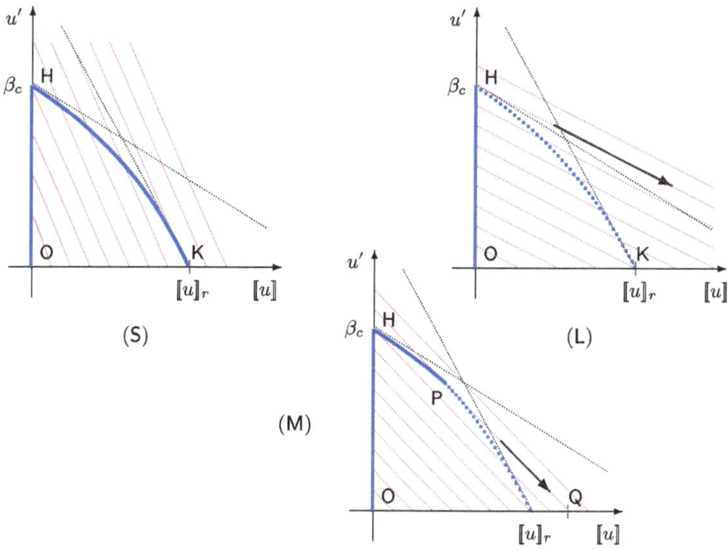

Fig. 4 Barenblatt's model for $((w')^{-1} \circ \theta')$ strictly concave in $(0, [\![u]\!]_r)$. The equilibrium curves (*bold*), and the loading lines (*light*) for small (S), medium (M), and large l (L). The stable and unstable portions of equilibrium curves are represented with *solid lines* and *dotted lines*, respectively. *The arrows* show the non-equilibrium transitions occurring at partially brittle (M) and at totally brittle fracture (L)

Thus, all unfractured equilibrium configurations with $\beta < \beta_c$ are stable, and a partially fractured configuration is stable only if inequality (2.11) holds. Moreover, if the cohesive energy is strictly concave, a pre-fractured configuration is stable only if $\#u \leq 1$.

If the cohesive energy is concave, a sufficient condition for stability is that $\#u = 1$ and the strict inequality (2.14) holds. A sufficient condition for non-concave cohesive energies is given in [63, Sect. 3.3].

2.4 Equilibrium Curves

By Proposition 2.2, for θ strictly concave in $(0, [\![u]\!]_r)$ the only stable pre-fractured equilibrium configurations are those with $\#u \leq 1$. In the representation in the $(u', [\![u]\!])$ plane of Fig. 4, these configurations form two *equilibrium curves*. The curve ($\#u = 0$) of the unfractured configurations is the half-line $u' < \beta_c$ of the u'-axis, and the curve ($\#u = 1$) is the curve of equation

$$u' = \big((w')^{-1} \circ \theta'\big)([\![u]\!]), \tag{2.15}$$

as follows from (2.8) and from the constitutive equation (1.10). The boundary condition (1.22), which now has the form

$$u' + \frac{1}{l}[\![u]\!] = \beta, \tag{2.16}$$

is represented by a family of parallel lines, one for each β, called the *loading lines*. The equilibrium configurations corresponding to a given β are the intersections of the equilibrium curves with the loading line for β.

By Proposition 2.1, all configurations on the equilibrium curve $\#u = 0$ are stable. For the configurations on the curve $\#u = 1$, the slopes of the equilibrium curve and of the loading lines are obtained by differentiation of (2.15) and (2.16), respectively

$$\frac{du'}{d[\![u]\!]} = \frac{\theta''([\![u]\!])}{w''(u')}, \qquad \frac{du'}{d[\![u]\!]} = -\frac{1}{l}. \tag{2.17}$$

By condition (2.14), a configuration u is stable if the slope of the equilibrium curve at u is larger than the slope of the loading lines. Moreover, by the necessary condition (2.11), all configurations at which the slope of the equilibrium curve is less than the slope of the loading lines is unstable. The number of the stable equilibrium configurations corresponding to each β depends on the shape of the function $(w')^{-1} \circ \theta'$. The case of $(w')^{-1} \circ \theta'$ concave is shown in Fig. 4. The convex case will be considered later, see Fig. 8.

In Fig. 4, the tangents to the equilibrium curve (2.15) at $\mathsf{H} = (\beta_c, 0)$ and $\mathsf{K} = (0, [\![u]\!]_r)$ are shown. By (2.17), their slopes are

$$\left.\frac{du'}{d[\![u]\!]}\right|_{[\![u]\!]=0} = \frac{\theta''(0)}{w''(\beta_c)}, \qquad \left.\frac{du'}{d[\![u]\!]}\right|_{[\![u]\!]=[\![u]\!]_r} = \frac{\theta''([\![u]\!]_r)}{w''(0)}.$$

Because $-1/l$ is the slope of the loading line, these are the slopes of the loading line for

$$l = -\frac{w''(\beta_c)}{\theta''(0)} \doteq l_{\mathsf{ML}}, \qquad l = -\frac{w''(0)}{\theta''([\![u]\!]_r)} \doteq l_{\mathsf{SM}}, \tag{2.18}$$

respectively. The lengths l_{ML} and l_{SM} define the three classes

$$l < l_{\mathsf{SM}}, \qquad l_{\mathsf{SM}} < l < l_{\mathsf{ML}}, \qquad l > l_{\mathsf{ML}},$$

of the bars of *small, medium, large size*. They will be labeled S, M, L, respectively.

The figure shows that all configurations on the equilibrium curve $\#u = 1$ are stable in the case S and unstable in the case L. In the case M, there is only one of the loading lines which has a unique tangent to the equilibrium curve. This occurs at the point denoted by P. The portion of the curve on the left of P is stable, and the portion on the right is unstable.

For a given load process $t \mapsto \beta_t$, the corresponding quasi-static evolutions are given by the intersections of the equilibrium curves with the loading lines for the loads β_t. Let us follow the evolution of the equilibrium configurations under a monotonically increasing load process starting from the *natural configuration* $(u', [\![u]\!]) = (0, 0)$ at $\beta = 0$.

Figure 4 shows that, in the case S, the loading line initially intersects the equilibrium curve $\#u = 0$. When β reaches the value β_c, the loading line starts intersecting the curve $\#u = 1$. This occurs at the point H, which marks the transition between the unfractured and the partially fractured regime. Total fracture is reached at the point $\mathsf{K} = (0, [\![u]\!]_r)$.

In the case M, the evolution is initially the same: a pre-fracture point is created when $\beta = \beta_c$, and for $\beta > \beta_c$ the evolution follows the curve $\#u = 1$. At the point P at which the loading line becomes tangent to the equilibrium curve, for further increasing β the only equilibrium configurations corresponding to β are the totally fractured configurations on the axis $u' = 0$. Then, a non-equilibrium transition occurs at constant β. In the figure, this corresponds to the transition from P to Q along the loading line.

In the case L, the only equilibrium configurations corresponding to $\beta > \beta_c$ are the totally fractured configurations. Then, at H there is a sharp non-equilibrium transition from the unfractured to a totally fractured configuration, without any intermediate regime of pre-fracture.

Thus, the classes S, M, L exhibit three different fracture modes. They correspond to the brittle, ductile-brittle, and ductile modes described in Sect. 1.2 and shown in Fig. 1.

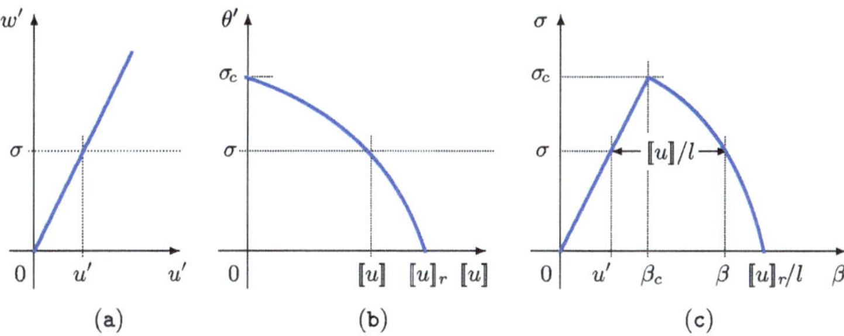

Fig. 5 Construction by points of the response curve (σ, β) for Barenblatt's model

2.5 Response Curves

Consider again a strictly concave θ. To each equilibrium curve in the $(u', [\![u]\!])$ plane there corresponds a *response curve* in the (σ, β) plane. For the unfractured equilibrium configurations u_β, this is the curve $\sigma = w'(\beta)$. For the partially fractured equilibrium configurations with $\#u = 1$, the equation of the response curve is obtained by solving the system (1.10), (2.8), (2.16)

$$\sigma = w'(u') = \theta'([\![u]\!]), \qquad \beta = u' + \frac{1}{l}[\![u]\!].$$

There is a correlation between stability and the sign of the slope of the response curve. Indeed, if we consider u' and $[\![u]\!]$ as functions of β, by differentiation we get

$$\frac{d\sigma}{d\beta} = w''(u')\frac{du'}{d\beta} = \theta''([\![u]\!])\frac{d[\![u]\!]}{d\beta}, \qquad 1 = \frac{du'}{d\beta} + \frac{1}{l}\frac{d[\![u]\!]}{d\beta}, \tag{2.19}$$

and, after elimination of $du'/d\beta$ and $d[\![u]\!]/d\beta$,

$$\frac{d\sigma}{d\beta} = \frac{\theta''([\![u]\!])w''(u')}{\theta''([\![u]\!]) + l^{-1}w''(u')}. \tag{2.20}$$

On the right side, by the stability conditions (2.11), (2.14), the denominator is positive for stable configurations and negative for unstable configurations. Moreover, the numerator is negative because θ is concave and w is convex. Then the slope $d\sigma/d\beta$ of the response curve for partially fractured configurations with $\#u = 1$ is negative at the stable points of the curve, and positive at the unstable points.

A graphical construction of the response curve for pre-fractured configurations is shown in Fig. 5. For each σ, the equilibrium values of u' and $[\![u]\!]$ are given by the graphs of $w'(u')$ and $\theta'([\![u]\!])$ in Figs. 5(a) and 5(b). The point (σ, β) of the response curve in Fig. 5(c) is obtained by summing u' and $[\![u]\!]/l$, as required by (2.16).

The shape of the response curve depends on the length l of the bar. This dependence is shown in Fig. 6(a), where the response curves for three bar lengths $l_1 < l_2 < l_3$, one for each of the classes S, M, L, are given. The light dotted lines are the response curves for the lengths l_{SM} and l_{ML}, defined in (2.18), which separate the three classes S, M, L. They are the lines with infinite slope at $\sigma = 0$ and at $\sigma = \sigma_c$, respectively.

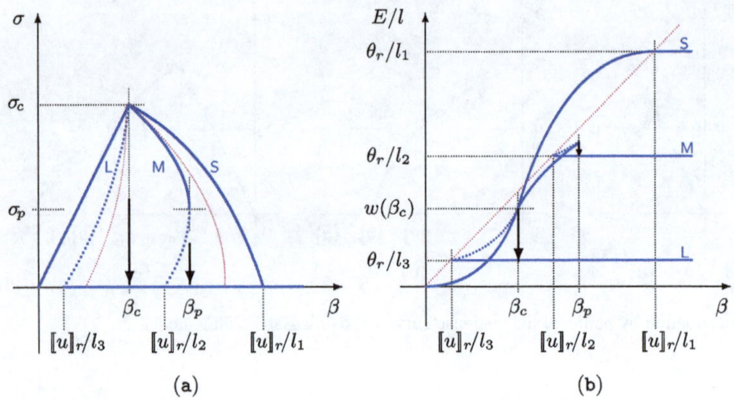

Fig. 6 Force-elongation (**a**) and energy-elongation (**b**) response curves for small (S), medium (M), and large-size bars (L). *The dotted lines* denote unstable equilibrium branches. *The arrows* show the non-equilibrium transitions at total fracture (**a**), and the corresponding energy dissipations (**b**)

For all l, the evolution follows the curve $\sigma = w'(\beta)$ up to the point (σ_c, β_c), at which a crack opens and the axial force reaches its maximum $\sigma_c = w'(\beta_c) = \theta'(0)$. For $\beta > \beta_c$ the shape of the response curve depends on l. For $l = l_1$, the whole curve has a negative slope. The force gradually decreases, and vanishes at the point $(0, [\![u]\!]_r/l_1)$ at which total fracture occurs. For $l = l_2$ the slope is negative up to a point (σ_p, β_p), at which it switches from $-\infty$ to $+\infty$. The sharp non-equilibrium transition to the line of the totally fractured configurations is marked by an arrow. For $l = l_3$ the slope of the whole curve is positive, and the non-equilibrium transition occurs just after crack opening.

The energy-elongation response curves are shown in Fig. 6(b), where, for the same lengths l_1, l_2, l_3 as above, the ratio E/l is represented as a function of β. By differentiation of the relation

$$\frac{1}{l}E(u) = w(u') + \frac{1}{l}\theta([\![u]\!]), \tag{2.21}$$

from (2.8) and $(2.19)_2$ it follows that

$$\frac{1}{l}\frac{dE}{d\beta} = w'(u')\frac{du'}{d\beta} + \frac{1}{l}\theta'([\![u]\!])\frac{d[\![u]\!]}{d\beta} = \sigma\left(\frac{du'}{d\beta} + \frac{1}{l}\frac{d[\![u]\!]}{d\beta}\right) = \sigma. \tag{2.22}$$

That is, the (σ, β) curve is the derivative of the $(E/l, \beta)$ curve. Therefore, the ordinates in Fig. 6(b) correspond to areas in Fig. 6(a).

The portion $\beta < \beta_c$ of the $(E/l, \beta)$ curve is the same for all l, while for $\beta > \beta_c$ the shape of the curve depends on the length of the bar. For $l = l_1$ the energy E/l increases, but at a slower rate than in the unfractured regime. At total fracture, $\beta = [\![u]\!]_r/l_1$, the response curve has a horizontal tangent, and there is a smooth transition to the curve $E/l = \theta_r/l_1$ of the totally fractured configurations.

For $l = l_2$, E/l increases at an even slower rate, and total fracture takes place at $\beta = \beta_p$. The unstable branch continues backward, up to complete unloading, $\sigma = 0$, at $\beta = [\![u]\!]_r/l_2$. The non-equilibrium transition in Fig. 6(a) corresponds to a sharp decrease of energy at $\beta = \beta_p$, where E/l drops to the energy θ_r/l_2 of the totally fractured configuration.

For $l = l_3$, after total fracture at $\beta = \beta_c$, the backward continuation up to complete unloading at $\beta = [\![u]\!]_r / l_3$ is unstable. In the non-equilibrium transition at $\beta = \beta_c$, the energy drops from $l_3 w(\beta_c)$ to θ_r. The energy $(l_3 w(\beta_c) - \theta_r)$ is dissipated. Therefore, the energy dissipated at total fracture is included between zero in the case S and $(l w(\beta_c) - \theta_r)$ in the case L.

This analysis reveals the dissipative character of the non-equilibrium transitions. A more accurate investigation of the dissipative phenomena associated with the cohesive energy will be made later, in Lecture 4.

2.6 Energy Landscapes

In this Section we give a specific example of the evolution of an energy of the form (2.21), under a monotonic increasing load β. In (2.21), take

$$l = 1, \qquad w(u') = \frac{1}{4}u'^2, \qquad \theta([\![u]\!]) = \begin{cases} [\![u]\!] - \frac{1}{4}[\![u]\!]^4 & \text{if } 0 \le [\![u]\!] \le 1, \\ \frac{3}{4} & \text{if } [\![u]\!] > 1. \end{cases} \tag{2.23}$$

The cohesive energy θ is non-decreasing, concave, and satisfies the assumptions (2.1). Moreover, θ is C^2 everywhere on the positive half-line, since the function and the first and second derivatives are continuous at $[\![u]\!] = 1$. The value of β and u' at crack opening, the value of $[\![u]\!]$ at total fracture, and the rupture energy are

$$\beta_c = 2, \qquad [\![u]\!]_r = 1, \qquad \theta_r = \frac{3}{4}, \tag{2.24}$$

respectively. For a pre-fractured configuration, the total energy (2.21) is

$$E(u) = \frac{1}{4}u'^2 + [\![u]\!] - \frac{1}{4}[\![u]\!]^4, \quad 0 \le [\![u]\!] \le 1. \tag{2.25}$$

This is the energy of a medium-size bar, case (b) of Fig. 4(b). Indeed, since

$$w''(u') = \frac{1}{2}, \qquad \theta''([\![u]\!]) = -3[\![u]\!]^2, \tag{2.26}$$

from the definitions (2.18) we have $l_{\mathrm{ML}} = +\infty$ and $l_{\mathrm{SM}} = 1/6$, and $l = 1$ is in between.

In Fig. 7(a) the equilibrium curves and the loading lines are shown. The equilibrium curve for $\#u = 0$ is the half-line $u' < 2$ of the u'-axis, and the equilibrium curve (2.15) for $\#u = 1$ and the loading lines (2.16) are

$$u' = 2(1 - [\![u]\!]^3), \qquad u' = \beta - [\![u]\!], \tag{2.27}$$

respectively. In the same figure, the energy $E(u)$ is represented by level curves. The energy levels are the multiples of 0.125 in the interval (0, 1.5), and the loading lines refer to the loads

$$\beta_k = k^{1/2}, \quad k \in \{1, \dots 6\}.$$

The figure shows that the loading lines for β_1 to β_4 intersect both equilibrium curves, the line for β_5 only intersects the curve $\#u = 1$ at the point P, and the line for β_6 has no intersection

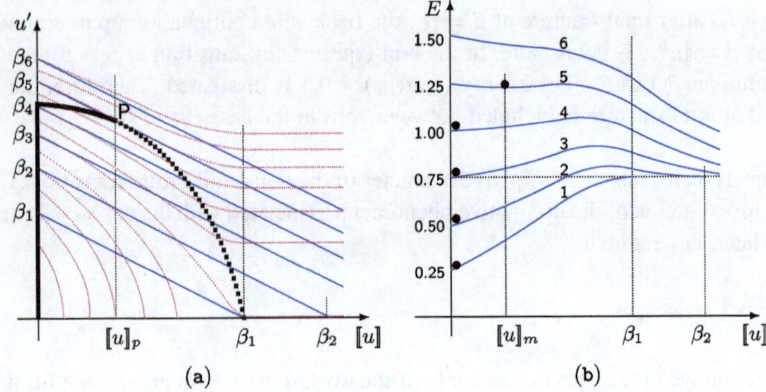

Fig. 7 Barenblatt's model. Level curves (**a**) and energy profiles on selected loading lines (**b**), for the strictly concave energy (2.23) of a medium-size bar. In (**a**), *the thick line* is the equilibrium curve. Its *solid part* is stable, and the *dotted part* is unstable. In (**b**), *the bullets* mark the positions of the local minima

with the equilibrium curves. The value $[\![u]\!]_p$ of $[\![u]\!]$ at P is the value at which the stability condition (2.11) is verified in the limit, as an equality. From (2.26),

$$0 = l\theta''([\![u]\!]_p) + w''(u'_p) = -3[\![u]\!]_p^2 + \frac{1}{2},$$

we get $[\![u]\!]_p = 1/\sqrt{6} \approx 0.408$.

Due to the relation $u' = \beta - [\![u]\!]$, for every fixed β the energy (2.25) can be expressed as a function of the single variable $[\![u]\!]$

$$E_\beta([\![u]\!]) = \begin{cases} \frac{1}{4}(\beta - [\![u]\!])^2 + [\![u]\!] - \frac{1}{4}[\![u]\!]^4 & \text{if } 0 \le [\![u]\!] \le 1, \\ \frac{1}{4}(\beta - [\![u]\!])^2 + \frac{3}{4} & \text{if } [\![u]\!] > 1. \end{cases} \tag{2.28}$$

For each β, this equation gives the energy profile at the corresponding loading line. The energy profiles for the selected loads β_k are shown in Fig. 7(b). For each β_k, the energy at $[\![u]\!] = 0$

$$E_{\beta_k}(0) = \frac{1}{4}\beta_k^2 = \frac{1}{4}k$$

is marked on the vertical axis.

At the partially fractured configurations, $0 < [\![u]\!] \le 1$, the derivative of the energy is

$$E'_\beta([\![u]\!]) = \frac{1}{2}(\varphi([\![u]\!]) - \beta), \qquad \varphi([\![u]\!]) = 2 + [\![u]\!] - 2[\![u]\!]^3.$$

As shown in Fig. 8, the minimum of φ is 1 at $[\![u]\!] = 1$, and the maximum is $2 + 2/\sqrt{54} \approx 2.272$ at $[\![u]\!] = 1/\sqrt{6} \approx 0.408$. Therefore, for $\beta < 1$ the energy is monotonic increasing. In Fig. 7(b), the energy profile for $\beta = \beta_1 = 1$ shows the limit situation in which the curve still has a maximum at $[\![u]\!] = 1$.

For $\beta > 1$, a maximum appears at some $[\![u]\!] < 1$. Indeed, for $\beta > 1$ the curve $\varphi = \varphi([\![u]\!])$ of Fig. 8 intersects the line $\varphi = \beta$, and the position $[\![u]\!]_k$ of the intersection corresponds to the maximum of the energy. This is the case of the energy profiles for β between β_2 and β_5 in Fig. 7(b), from which we see that, with increasing β, $[\![u]\!]_k$ tends to the limit value 0.408.

Fig. 8 Graph of the function φ for the energy (2.28). *The gray area* is the height of the energy barrier preventing the transition from pre-fractured to totally fractured configurations

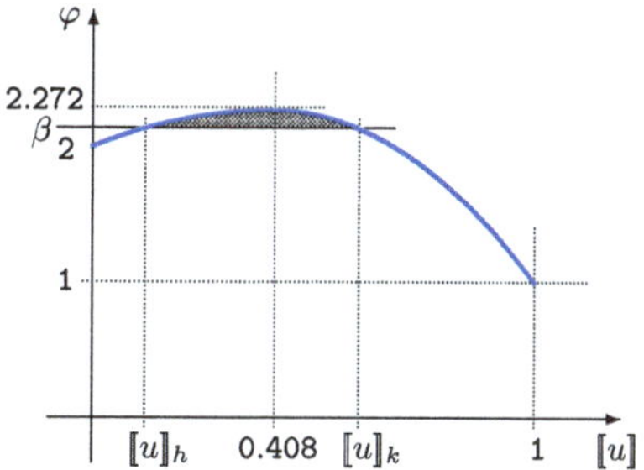

The same energy profiles show that the equilibrium configurations located on the right of the point P of Fig. 7(a) correspond to energy maxima and, therefore, to unstable equilibrium configurations.

For $\beta > 2$, a second intersection appears at some $[\![u]\!]_h < 0.408$. It corresponds to an energy minimum. The load $\beta = 2$ marks the transition between the elastic regime, in which the energy minimum is at the unfractured configuration, $[\![u]\!] = 0$, and the inelastic regime, in which the energy minimum is at a partially fractured configuration with $[\![u]\!] = [\![u]\!]_h$. In Fig. 7(b), this is the situation shown by the energy profile $\beta = \beta_5$. In fact, $\beta_5 = \sqrt{5} \approx 2.236$ is slightly less than the limit value $\beta = 2.272$ at which both intersections $[\![u]\!]_h$ and $[\![u]\!]_k$ disappear.

For $\beta > 2.272$ the energy $E_\beta([\![u]\!])$ is monotonic decreasing, and there are no equilibrium configurations. This is the situation for the energy profile $\beta = \beta_6$ in Fig. 7(b).

For each β, the difference between the maximum and the minimum of E_β is the height of the energy barrier preventing the transition from a pre-fractured to totally fractured configurations. From the equality

$$E_\beta([\![u]\!]_k) - E_\beta([\![u]\!]_h) = \frac{1}{2} \int_{[\![u]\!]_h}^{[\![u]\!]_k} (\varphi([\![u]\!]) - \beta) d[\![u]\!],$$

we see that the height of the energy barrier is measured by the gray area in Fig. 8. This area decreases with increasing β, and vanishes for $\beta > 2.272$.

2.7 References and Comments

In the 1950's, an intense research activity was addressed to understanding the physical mechanisms of brittle fracture. The reason was to find an explanation for the many catastrophic failures suffered by welded iron ships during the Pacific War [75]. At that time, the essential role played by plastic deformation in mitigating the brittle character of fracture was fully recognized. The substantial progress made by Orowan [109], Irwin [85], and others, substantially consisted in introducing the plastic work in the energy balance governing the crack growth. The constant bridging force in Dugdale's cohesive energy originates from the assumption of a perfectly plastic response at the crack tip.

Barenblatt's model has a different origin. The cohesive energy is supposed to be due to interatomic interaction [9]. Its concave shape comes from the Lennard-Jones potential, restricted to macroscopically observable distances [26, 128]. In a sense, Barenblatt's model is an updated version of some classical theories of the structure of matter, such as Boscovich's interaction law between particles [19] and Coulomb's theory of rupture [44], formulated long before the present level of knowledge of microscopic phenomena was reached.

Following Barenblatt's ideas, the cohesive energy model, mostly in Hillerborg's version of the *fictitious crack model* [81, 82], was soon applied to the fracture of concrete and other non-crystalline materials. To the writer's knowledge, it is within this research trend that, for the first time, it was explicitly stated that fracture is in fact a special form of instability of equilibrium [13, 31, 32, 108].

In the attempts of formulating a mathematical theory of fracture, the nasty properties of Griffith's energy functional (1.19) suggested the adoption of various forms of regularization. In [6, 23] the fracture energy was approximated by an integral involving a scalar field with values in [0, 1], measuring the *intensity* of fracture. More or less at the same time, fracture energies depending on the jump amplitudes were used to formulate singular perturbation problems [20, 27, 28]. A basic difference with the cohesive crack model already developed in fracture mechanics is that, while the latter is a self-consistent, physically motivated model, the regularized models of the mathematicians were regarded as approximations, in the sense of Γ-convergence, of the Griffith model. Griffith's theory was considered as the best mathematical representation of fracture. In fact, as explained in this second Lecture, in the one-dimensional case this is correct only for the bars of the class L, for which the fracture is totally brittle.

In the variational approach to fracture, the advantages of the cohesive energy models over Griffith's approach soon became evident [41, 63, 102]. Essentially, there are three main advantages. The first is that the cohesive energy models solve the problem of the characterization of crack initiation. The second is the possibility of a unified description of brittle, ductile-brittle, and ductile fracture, and the third is a precise explanation of the size effect. Other relevant aspects of the cohesive energy model are briefly commented below.

1. The cohesive energy model provides a sound mathematical basis for the problem of fracture. In particular, the definition of the distance (1.30) between configurations gives a precise characterization of the local energy minima. Notice that (1.30) is a true distance only for pre-fractured configurations. It becomes a pseudo-distance, like in Griffith's model, if totally fractured configurations are considered.

2. For a concave cohesive energy, the relevant configurations are those with constant u' and with at most one jump point. Because they are defined by the two numbers u' and $[\![u]\!]$, these configurations form a two-dimensional subspace of the configuration space. In this subspace, as in every finite-dimensional space, all metrics are equivalent. In particular, the Euclidean metric is equivalent to the metric (1.30). This makes legitimate the identification of the local minima in the sense of (1.30) with the local minima of the energy surface in Fig. 7.

3. In a cohesive energy model, any two pre-fractured configurations which can be joined by an equilibrium path are accessible from each other. Indeed, of the two conditions for accessibility given in Sect. 1.5, the condition of the irreversibility of fracture does not apply to pre-fractured configurations. For the second condition, for any equilibrium path from (2.22) we have

$$\dot{E}_t = \frac{dE}{d\beta}\dot{\beta}_t = l\sigma_t\dot{\beta}_t, \tag{2.29}$$

and therefore the energy inequality (1.33) is satisfied as an equality. That is, there is no dissipation in a deformation process involving pre-fractured configurations.

For example, the pre-fractured configurations O and P in Fig. 4(b) are accessible from each other, because they are joined by the equilibrium path (OHP). When the path is traversed backwards, the fracture present at P disappears at H. On the contrary, P is not accessible from the totally fractured configuration Q, because Q has at least one totally fractured point, and at this point the fracture is irreversible.

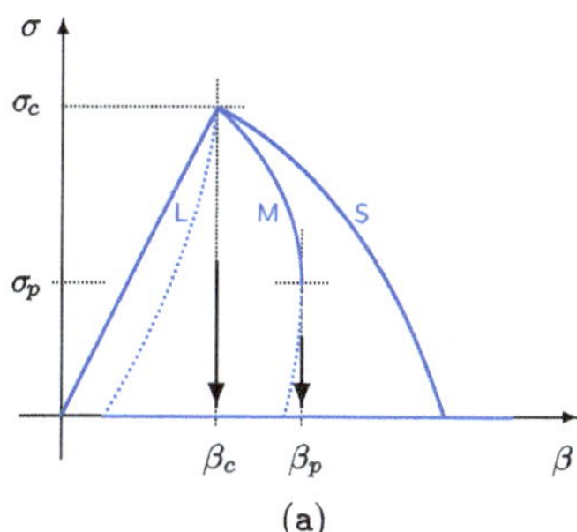

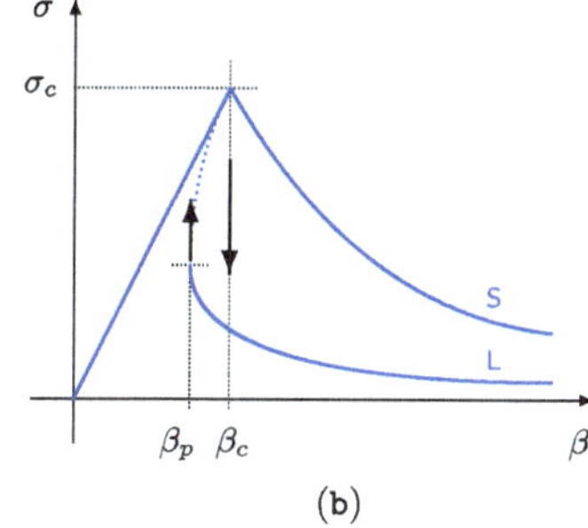

Fig. 9 Comparison of the (σ, β) response curves for bars of different sizes, for θ concave and $((w')^{-1} \circ \theta')$ concave (**a**) and convex (**b**). *The dotted lines* denote unstable configurations, and *the arrows* denote non-equilibrium transitions

4. Figures 4 to 6 refer to a concave energy θ with $((w')^{-1} \circ \theta')$ concave. The response curves of Fig. 6(a) are reproduced in Fig. 9(a), to be compared with those for θ concave and $((w')^{-1} \circ \theta')$ convex. As shown in Fig. 9(b), in the convex case there are only two classes, S and L, of bar lengths. After the onset of fracture at (σ_c, β_c), in case S the response curve continues with a negative slope. In case L it continues with a positive slope, with β decreasing up to a minimum $\beta = \beta_p$, where the slope becomes infinite and switches from $+\infty$ to $-\infty$. By the relation (2.20) between stability and the slope of the response curve, the first branch is unstable and the second branch is stable.

For bars of the class L, a non-equilibrium transition occurs at $\beta = \beta_c$. If, as in the case shown in the figure, the curve has a horizontal asymptote at $\sigma = 0$, then there is no total fracture. For $\beta > \beta_c$ the only existing equilibrium configurations are the pre-fractured configurations on the lower branch of the curve.

On the contrary, if $[\![u]\!]_r$ is finite, for $\beta > \beta_c$ there are both pre-fractured and totally fractured equilibrium configurations, and the model does not say at which of them the non-equilibrium transition ends. The assumption of $[\![u]\!]_r$ finite is physically more plausible. In fact, in Barenblatt's original formulation [9], the assumption of $[\![u]\!]_r$ finite and conveniently small was explicitly made.

5. Since the formulation of the fictitious crack model it was realized that, while a concave $((w')^{-1} \circ \theta')$ reproduces the response of metals, a convex $((w')^{-1} \circ \theta')$ is appropriate to concrete, both ordinary and fiber-reinforced, to rocks, fiber-reinforced plastics, bricks, and wood [81, 82]. This is largely confirmed by experiments, see, e.g., [83, 99, 126].

6. To Irwin [86] is due the idea that a cohesive energy defines an *internal length* of the material. In the Barenblatt-Dugdale model with a concave $((w')^{-1} \circ \theta')$, *two* internal lengths, l_{SM} and l_{ML}, naturally arise. They separate the three classes S, M, L of bar lengths, for which the fracture is ductile, ductile-brittle, and brittle, respectively. For a convex $((w')^{-1} \circ \theta')$ the class M disappears, and only the internal length separating the classes S and L remains.

7. We conclude with a short comment on the size effect. In general, the term *scaling* denotes a change in structural response due to a change of physical dimension. Scaling effects are frequent in both solid and fluid mechanics [10]. A particular scaling effect is the *size effect* of fracture mechanics, which is the property that large-size structures are more brittle than small-size structures made of the same material [11, 16, 33, 132]. This property was known for a long time to builders and craftsmen, though its underevaluation was the cause of several structural inconveniences over the centuries. The size effect was understood and explicitly stated for the first time by Galilei [72]. An outlook to earlier statements and successive historical developments can be found in the review paper [14].

3 Lecture 3. Microstructures

3.1 Convex-Concave Energies

In the preceding Lecture attention was focused on concave energies, which proved to be appropriate for a description of several phenomena related to fracture. Other interesting forms of material response can be captured by non-concave cohesive energies. In this Lecture, we first consider energies which are convex near the origin and concave everywhere else. Then, two other classes of non-concave energies, *bi-modal* and *periodic*, are briefly discussed.

Let θ be the energy represented in Fig. 10(a). Due to the presence of an inflection point at $[\![u]\!] = [\![u]\!]_h$, this energy is strictly convex in $(0, [\![u]\!]_h)$ and strictly concave in $([\![u]\!]_h, [\![u]\!]_r)$:

$$\theta''([\![u]\!]) > 0 \quad \text{if } 0 < [\![u]\!] < [\![u]\!]_h, \qquad \theta''([\![u]\!]) < 0 \quad \text{if } [\![u]\!]_h < [\![u]\!] < [\![u]\!]_r.$$

As shown in Fig. 10(b), a consequence of the initial convexity is that there is a crack amplitude $[\![u]\!]_k > [\![u]\!]_h$ such that

$$\theta'([\![u]\!]) > \theta'(0) \quad \forall [\![u]\!] < [\![u]\!]_k, \tag{3.1}$$

and, by consequence, the equilibrium condition (2.7) is violated at all $[\![u]\!]$ in $(0, [\![u]\!]_k)$. That is, in a pre-fractured equilibrium configuration all jump amplitudes must be greater or equal than $[\![u]\!]_k$. Moreover, as a consequence of the strict concavity of θ in $([\![u]\!]_h, [\![u]\!]_r)$, all equilibrium configurations with $\#u > 1$ are unstable by Proposition 2.2. Therefore,

$$\#u = 1 \quad \text{and} \quad [\![u]\!] \geq [\![u]\!]_k \tag{3.2}$$

are necessary conditions for the stability of a pre-fractured equilibrium configuration.

All stable pre-fractured configurations lie in the plane $(u', [\![u]\!])$. In this plane, all partially fractured equilibrium configurations are contained in the half-plane $[\![u]\!] \geq [\![u]\!]_k$, which is separated from the line $[\![u]\!] = 0$ of the unfractured configurations by the positive distance $[\![u]\!]_k$. This separation renders impossible a smooth transition from the unfractured to the partially fractured configurations or vice versa. That is, crack opening can only be made with a non-equilibrium transition. In particular, for small-size bars, to each β corresponds at most one pre-fractured equilibrium configuration. Therefore, the only possible evolution is the catastrophic transition to a totally fractured configuration. Before accepting this rather extreme conclusion, let us explore the possibility that a smooth transition be obtained by introducing some sort of *generalized configurations*.

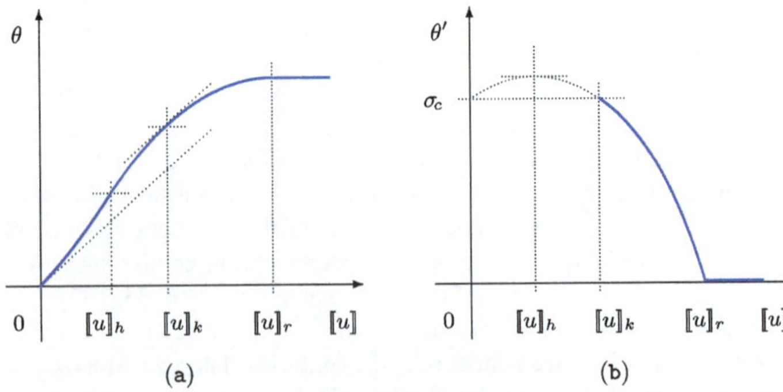

Fig. 10 A convex-concave cohesive energy (**a**), and its derivative (**b**)

3.2 Generalized Configurations

Let $f:[0,l]\to\mathbb{R}$ be a differentiable function, strictly increasing, and with

$$f(0)=0, \qquad f(l)=\beta l, \tag{3.3}$$

and let $g:[0,l]\to\mathbb{R}$ be a continuous function such that

$$f'(x)\geq g(x)>0 \quad \forall x\in(0,l). \tag{3.4}$$

For each n, select $n+1$ points x_n^i such that

$$x_n^0=0, \qquad x_n^{i-1}<x_n^i\leq x_n^{i-1}+\frac{c}{n}, \qquad i=1,\dots n, \qquad x_n^n=l, \tag{3.5}$$

for some constant $c\geq 1$. Then for every x in $(0,l)$ there is a sequence of integers i_n such that x belongs to $[x_n^{i_n-1},x_n^{i_n})$ and

$$0\leq x-x_n^{i_n-1}<c/n. \tag{3.6}$$

Consider the sequence

$$u_n(x)=f\left(x_n^{i_n-1}\right)+\int_{x_n^{i_n-1}}^{x} g(\xi)d\xi \quad \forall x\in\left(x_n^{i_n-1},x_n^{i_n}\right). \tag{3.7}$$

For it,

$$f(x)-u_n(x)=\int_{x_n^{i_n-1}}^{x}\left(f'(\xi)-g(\xi)\right)d\xi.$$

The difference $(f(x)-u_n(x))$ is positive by the positiveness of $(f'-g)$. Moreover, the lengths of the intervals (x_n^{i-1},x_n^i) converge to zero uniformly because, by (3.5), they are all less or equal than c/n. Then $(f(x)-u_n(x))$ tends to zero uniformly by the uniform continuity of the integral. That is,

$$\lim_{n\to\infty}\sup_{x\in(0,l)}|f(x)-u_n(x)|=0. \tag{3.8}$$

Every function u_n has n jumps, one at every point x_n^i, with jump amplitude

$$[\![u_n]\!](x_n^i)=f(x_n^{i_n})-f(x_n^{i_n-1})-\int_{x_n^{i_n-1}}^{x_n^i} g(x)dx. \tag{3.9}$$

Then, summing over i,

$$\sum_i^n [\![u_n]\!](x_n^i)=f(l)-f(0)-\int_{x_n^{i_n-1}}^{x_n^i} g(x)dx=\int_0^l\left(f'(x)-g(x)\right)dx. \tag{3.10}$$

Though the jumps diffuse over the bar and, in the limit, become macroscopically invisible, the sum of their amplitudes keeps the constant value given by the integral of the difference $f'-g$. Writing the total elongation $\beta l=f(l)-f(0)$ in the form

$$\beta l=\int_0^l\left(f'(x)-g(x)\right)dx+\int_0^l g(x)dx, \tag{3.11}$$

one sees that βl is the sum of the total elongation (3.10) due to the jumps plus the integral of g, which is the total elongation due to the bulk deformation $u_n' = g$.

A pair (f, g) satisfying conditions (3.3) and (3.4) is a *generalized configuration*, or *configuration with microstructure*. The function f is the *macroscopic configuration*, and g is the *elastic deformation* of (f, g). We say that (f, g) is a *microstructure corresponding to* β if f is a configuration corresponding to β.

The difference $f'(x) - g(x)$ is a measure of the quantity of microstructure at x. To make this idea precise, consider the expression (3.9) of the jump amplitudes. Dividing by $(x_n^i - x_n^{i-1})$, in the limit we get

$$\lim_{n\to\infty} \frac{[\![u_n]\!](x_n^i)}{x_n^i - x_n^{i-1}} = f'(x) - g(x). \tag{3.12}$$

This shows that the difference $f'(x) - g(x)$ is the amount of microscopic jumps which, in the limit, concentrates at x. We call it the *density of microstructure* at x. In particular, the limit case $f' = g$ corresponds to a configuration without microstructure. That is, the generalized configuration (f, f') coincides with the ordinary configuration represented by the function f.

A sequence $n \mapsto u_n$ with $u_n' = g$ and with the convergence property (3.8) is an approximating sequence for (f, g). More in general, an *approximating sequence for* (f, g) is a sequence $n \mapsto u_n$ such that

$$\lim_{n\to\infty} \sup_{x\in(0,l)} |f(x) - u_n(x)| = 0, \qquad \lim_{n\to\infty} \sup_{x\in(0,l)} |g(x) - u_n'(x)| = 0. \tag{3.13}$$

That is, the derivative u_n', instead of being equal to g, converges uniformly to g. The distance (1.30) can be extended to the generalized configurations by setting

$$\|(f, g) - (f_o, g_o)\| = \sup_{x\in(0,l)} |f(x) - f_o(x)| + \sup_{x\in(0,l)} |g(x) - g_o(x)|. \tag{3.14}$$

In particular, if (f_o, g_o) is a configuration without microstructure, that is, if it is an ordinary configuration (u, u'), the distance is

$$\|(f, g) - u\| = \sup_{x\in(0,l)} |f(x) - u(x)| + \sup_{x\in(0,l)} |g(x) - u'(x)|. \tag{3.15}$$

Then the convergence (3.13) of the approximating sequences is in fact convergence in the set of all generalized configurations, equipped with the metric (3.14).

The energy of a microstructure can be defined as the limit of the energies of the approximating sequences. This definition makes sense only if the limit is the same for all approximating sequences. To show that this is the case, take an approximating sequence $n \mapsto u_n$ for (f, g). Each u_n has the energy

$$E(u_n) = \int_0^l w(u_n'(x))dx + \sum_{i=1}^n \theta([\![u_n]\!](x_n^i)).$$

By the continuity of w and the uniform convergence of u_n' to g, the integral term converges to the integral of $w(g(x))$. For the remaining term, from (3.12) we have that $[\![u_n]\!](x_n^i)$ goes to zero as fast as $(x_n^i - x_n^{i-1})$, that is, as fast as $1/n$. Therefore,

$$\theta([\![u_n]\!](x_n^i)) = \theta'(0)([\![u_n]\!](x_n^i)) + o(1/n),$$

and, by (3.10),

$$\lim_{n\to\infty}\sum_{i=1}^{n}\theta(\llbracket u_n\rrbracket(x_n^i)) = \lim_{n\to\infty}\left(\theta'(0)\sum_{i=1}^{n}(\llbracket u_n\rrbracket(x_n^i)) + n\, o(1/n)\right)$$
$$= \theta'(0)\int_0^l \big(f'(x)-g(x)\big)dx.$$

Then the energies $E(u_n)$ converge to the limit

$$\lim_{n\to\infty} E(u_n) = \int_0^l w\big(g(x)\big)dx + \theta'(0)\int_0^l \big(f'(x)-g(x)\big)dx.$$

Since this limit does not depend on the approximating sequence, it makes sense to define the energy of (f, g) as the limit of the energies of the approximating sequences. By (3.11), this limit can be given the form

$$E(f,g) = \int_0^l \Big(w\big(g(x)\big) + \theta'(0)\big(\beta - g(x)\big)\Big)dx. \tag{3.16}$$

This has the remarkable consequence that the energy depends on f only through β. That is, *all microstructures (f, g) with the same elastic deformation g and corresponding to the same β have the same energy.*

3.3 Equilibrium Microstructures

We say that (f, g) is an *equilibrium microstructure* if the first variation of the energy

$$\delta E(f,g,\zeta,\eta) = \lim_{\varepsilon\to 0}\frac{1}{\varepsilon}\big(E(f+\varepsilon\zeta, g+\varepsilon\eta) - E(f,g)\big)$$

is non-negative for all perturbations (ζ,η) which preserve the length of the bar, and such that the perturbed configuration $(f+\zeta, g+\eta)$ is itself a microstructure. The condition that the perturbation preserves the length of the bar is expressed by the equality

$$\int_0^l \zeta'(x)dx = 0, \tag{3.17}$$

and the requirement that $(f+\zeta, g+\eta)$ be a microstructure is expressed by the conditions

$$f'(x)+\zeta'(x)-g(x) \ge \eta(x) > -g(x) \quad \forall x\in(0,l), \tag{3.18}$$

which follow from (3.4). Recalling the identity $\theta'(0) = w'(\beta_c)$, from (3.16) it follows that

$$\delta E(f,g,\zeta,\eta) = \lim_{\varepsilon\to 0}\frac{1}{\varepsilon}\int_0^l \Big(w\big(g(x)+\varepsilon\eta(x)\big) - w\big(g(x)\big) - \varepsilon\theta'(0)\eta(x)\Big)dx$$
$$= \int_0^l \Big(w'\big(g(x)\big) - w'(\beta_c)\Big)\eta(x)dx. \tag{3.19}$$

The first variation is independent of ζ. The reason is that the energy depends on f only through β, and β must be the same for the unperturbed and the perturbed configuration.

By (3.19), the first variation is zero if g is the constant function

$$g_c(x) = \beta_c. \tag{3.20}$$

This condition is also necessary for non-negativeness if $f' \neq g$. Indeed, from (3.17) and (3.18),

$$\int_0^l (f'(x) - g(x))dx \geq \int_0^l \eta(x)dx \geq -\int_0^l g(x)dx. \tag{3.21}$$

Then the admissible perturbations η include all η with a null integral. The right-hand side of (3.18) is non-negative for all such η only if g is a constant, $g(x) = g_c$. Then the first variation reduces to

$$\delta E(f, g, \zeta, \eta) = \big(w'(g_c) - w'(\beta_c)\big) \int_0^l \eta(x)dx. \tag{3.22}$$

If $f' \neq g$, the integral of $(f' - g)$ is positive. Then by (3.21) the integral of η may take any sign, and therefore g_c must be equal to β_c. This proves that condition (3.20) is necessary for equilibrium if $f' \neq g$. That is, *a pair* (f, g) *with* $f' \neq g$ *is an equilibrium microstructure if and only if* $g(x) = \beta_c$ *and* f *satisfies conditions* (3.3) *and* (3.4).

If $f' = g$, then (f, g) is an ordinary configuration with $f'(x) = g(x)$ and, because $g(x) = g_c$, f is the homogeneous deformation $f(x) = g_c x$. Moreover, since the integral of $(f' - g)$ is zero, the integral of η is non-positive by (3.21). Then from (3.22) it follows that $w'(g_c) \leq w'(\beta_c)$, that is, $g_c \leq \beta_c$. Thus, in the more general context of the microstructures, it is confirmed that there are equilibrium configurations without microstructure corresponding to β only if $\beta \leq \beta_c$, and that the only equilibrium configuration without microstructure corresponding to such β is the homogeneous deformation u_β.

In Sect. 3.1 it was shown that a continuous transition from the unfractured to the pre-fractured configurations is impossible, because there are no pre-fractured equilibrium configurations at a distance less than $[\![u]\!]_k$ from the unfractured equilibrium configurations. Now we know that there are equilibrium microstuctures (f, g) with $g = \beta_c$ and with f arbitrarily close to the terminal point u_{β_c} of the half-line of the unfractured equilibrium configurations. Therefore, at $\beta = \beta_c$ a continuous transition from the unfractured configurations to the microstructures is possible.

The equilibrium microstructures (f, g_c) corresponding to the same $\beta \geq \beta_c$ form a family μ_β of equilibrium microstructures, all with the same energy

$$E(\mu_\beta) = l\big(w(\beta_c) + \theta'(0)(\beta - \beta_c)\big), \quad \beta \geq \beta_c, \tag{3.23}$$

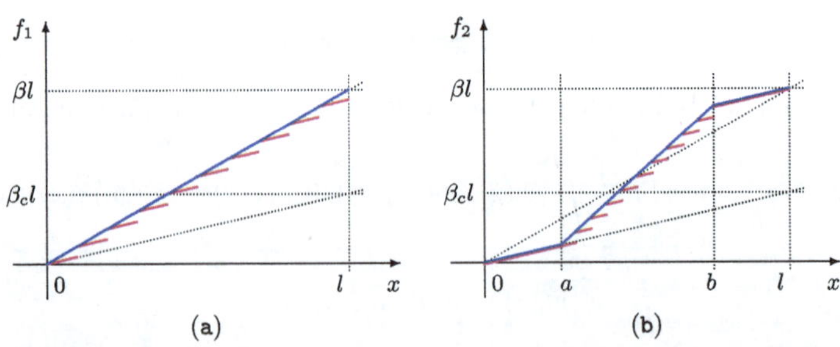

Fig. 11 Two equilibrium microstructures (f_1, g_c), (f_2, g_c) of the family μ_β, each with an element of an approximating sequence

and all with the same total amount of microstructure $l(\beta - \beta_c)$. This amount can be distributed over the bar in different ways, according to the punctual values of the density of microstructure $(f'(x) - \beta_c)$.

Two members f_1, f_2 of the family μ_β are shown in Fig. 11. The function f_1 in Fig. 11(a) is the linear function $f_1(x) = \beta x$. It has a uniform density of microstructure $(\beta - \beta_c)$. As shown in the figure, the microstructure (f_1, g_c) is the limit of an approximating sequence $n \mapsto u_n$ of pre-fractured configurations, in which $u_n' = \beta_c$ and each u_n has n equally spaced jumps of equal amplitude $l(\beta - \beta_c)/n$.

The function f_2 in Fig. 11(b) is a piecewise linear function with $f_2'(x) > \beta_c$ in the subinterval (a, b) and $f_2'(x) = \beta_c$ outside. The microstructure (f_2, g_c) is the limit of a sequence $n \mapsto u_n$, with $u_n' = \beta_c$ and with all jump points concentrated at (a, b), equally spaced and with equal jump amplitudes. The density of microstructure is uniform in (a, b), and zero outside.

3.4 Stable Microstructures

Let u be a pre-fractured configuration corresponding to $\beta \geq \beta_c$, with jump amplitudes less than $[\![u]\!]_k$. By the strict convexity of w, by the inequality

$$\theta([\![u]\!]) > \theta'(0)([\![u]\!]) \quad \forall [\![u]\!] \in (0, [\![u]\!]_k), \tag{3.24}$$

which follows from integration of (3.1) over $(0, [\![u]\!])$, by the identity $\theta'(0) = w'(\beta_c)$, and by the boundary condition (1.22),

$$\begin{aligned} E(u) &= \int_0^l w\big(u'(x)\big)dx + \sum_i \theta\big([\![u]\!](x_i)\big) \\ &> \int_0^l \big(w(\beta_c) + w'(\beta_c)\big(u'(x) - \beta_c\big)\big)dx + \theta'(0)\sum_i [\![u]\!](x_i) \\ &= l w(\beta_c) + \theta'(0)\bigg(\int_0^l \big(u'(x) - \beta_c\big)dx + \sum_i [\![u]\!](x_i)\bigg) \\ &= l w(\beta_c) + l\theta'(0)(\beta - \beta_c). \end{aligned} \tag{3.25}$$

By (3.23), the right side is the energy of the family μ_β. Then, the energy $E(\mu_\beta)$ is less than the energy of all pre-fractured configurations u corresponding to the same $\beta \geq \beta_c$ and with jump amplitudes less than $[\![u]\!]_k$.

Moreover, every (f, g_c) in μ_β is a local minimizer for all configurations u corresponding to β. Indeed, by the boundary condition (1.22) and by $(3.3)_2$,

$$\max_i [\![u]\!](x_i) \leq \sum_i [\![u]\!](x_i) = l\beta - \int_0^l u'(x)dx = f(l) - u(l),$$

and, for the distance (3.15),

$$\max_i [\![u]\!](x_i) \leq \sup_{x \in (0,l)} |f(x) - u(x)| \leq \|(f, g_c) - u\|.$$

Then all configurations u such that $\|(f, g_c) - u\| \leq [\![u]\!]_k$ have jump amplitudes not greater than $[\![u]\!]_k$. Therefore, by (3.25), their energy is greater than $E(f, g_c) = E(\mu_\beta)$.

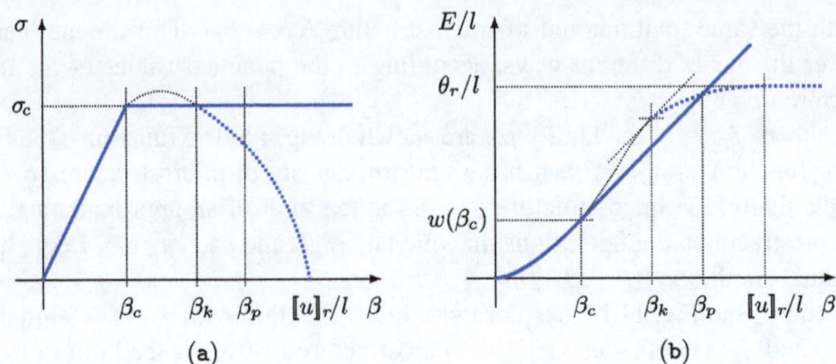

Fig. 12 Force-elongation (**a**) and energy-elongation (**b**) response curves for a convex-concave energy. The partially fractured configurations (*dotted curve*) are not accessible from the natural configuration

This minimum property can be extended to all microstructures (f, g) with f corresponding to β. Indeed, (f, g) is a limit of a sequence $n \mapsto u_n$ of pre-fractured configurations corresponding to β, each with jump amplitudes $[\![u_n]\!]$ converging to zero uniformly, and therefore with energy $E(u_n)$ greater than $E(f, g_c)$ for sufficiently large n. By definition, the limit of the energies $E(u_n)$ is the energy of (f, g). Therefore, by (3.18),

$$E(f, g) \geq E(f, g_c) = E(\mu_\beta). \tag{3.26}$$

That is, all microstructures (f, g_c) in the family μ_β are local energy minimizers in the extended class of all microstructures (f, g) with f corresponding to β.

3.5 Response Curves

The (σ, β) response curves for pre-fractured configurations can be constructed by points, with the same procedure used in Sect. 2.5 for the Barenblatt model. The response curves for small-size bars are shown in Fig. 12(a). The equilibrium curve for the unfractured configurations covers the interval $(0, \beta_c)$, and the curve for the partially fractured configurations covers the interval $(\beta_k, [\![u]\!]_r/l)$.

For an equilibrium microstructure (f, g_c), the axial force is defined as the limit of the axial forces in an approximating sequence $n \mapsto u_n$:

$$\sigma = \lim_{n\to\infty} w'\big(u_n'(x)\big) = w'(\beta_c).$$

Then, σ has the same value $\sigma_c = w'(\beta_c) = \theta'(0)$ for all equilibrium micro-structures. The corresponding response curve is the horizontal half-line $\sigma = \sigma_c$ starting at the point (σ_c, β_c). All equilibrium microstructures (f, g_c) have the same response curve, no matter which is the form of f. This curve fills the gap (β_c, β_k) at which there are no pre-fractured equilibrium configurations.

The energy-elongation response curve $(E/l, \beta)$ is shown in Fig. 12(b). Like in Barenblatt's model, for $\beta < \beta_c$ the response follows the curve $E/l = w(\beta)$ of the unfractured configurations. For $\beta > \beta_c$ there are two curves, the curve of the pre-fractured configurations for $\beta > \beta_k$ and the curve of the microstructures for $\beta > \beta_c$. With increasing β, the slope of the first curve decreases, and eventually becomes horizontal at total fracture, $\beta = [\![u]\!]_r/l$. The curve of the microstructures is a straight half-line, tangent to the curve $E/l = w(\beta)$ at the initial point $(w(\beta_c), \beta_c)$.

The two curves meet at some $\beta_p > \beta_k$. At this point, just as for crack initiation in Griffith's model, there is the alternative between continuation along the line of the microstructures, and the creation of a macroscopic jump, followed by continuation along the curve of the pre-fractured configurations, up to total fracture at $\beta = [\![u]\!]_r/l$.

The second possibility is excluded, because it requires a non-equilibrium transition from a microstructure to a pre-fractured configuration, and any such transition violates the energetic dissipation inequality (1.33). To see this, take a microstructure (f, g_c) of the family $\mu(\beta_p)$ and a pre-fractured configuration u_p corresponding to β_p, and let $[\![u]\!]_p$ be the largest amplitude of the jumps of u_p. Then,

$$\|(f, g_c) - u_p\| \geq \sup_{x\in(0,l)} |f(x) - u_p(x)| \geq \frac{1}{2}[\![u]\!]_p. \tag{3.27}$$

Indeed, if x_p is the point at which $[\![u_p]\!](x_p) = [\![u]\!]_p$ and if $f(x) - u_p(x-) = \delta$, then

$$f(x) - u_p(x+) = f(x) - u_p(x-) - [\![u_p]\!](x_p) = \delta - [\![u_p]\!],$$

and at least one among $|\delta|$ and $|\delta - [\![u_p]\!]|$ is greater or equal than $\frac{1}{2}[\![u]\!]_p$. In particular, if u_p is a stable equilibrium configuration, inequality (3.27) shows that the distance $\|(f, g_c) - u_p\|$ is finite. Because there are no other stable equilibrium configurations corresponding to β_p, the pre-fractured configuration u_p can be reached from (f, g_c) only with a non-equilibrium transition. Let the transition be made with a non-equilibrium process $t \mapsto u_t$. Then the distance $\|(f, g_c) - u_t\|$ starts from zero and reaches a final value which, by (3.27), is not smaller than $\frac{1}{2}[\![u]\!]_p$. Then, at some t, the distance $\|(f, g_c) - u_t\|$ reaches a positive value smaller than the value $[\![u]\!]_k$ shown in Fig. 10. For this t, by the strict inequality in (3.25),

$$E(u_t) - E(f, g_c) = a > 0.$$

That is, to reach u_p from (f, g_c) it is necessary to overcome a barrier of height a. Because β is constant over the process, no power is supplied from the exterior. Therefore, the energy inequality (1.33) is violated.

This leads to the conclusion that the equilibrium curve of the pre-fractured configurations is inaccessible from the curve of the equilibrium microstructures. Consequently, the only equilibrium configurations corresponding to $\beta > \beta_c$ which are accessible from the natural configuration are the microstructures of the family μ_β.

3.6 A Glance at Other Cohesive Energies

A comparison of the response curves in Fig. 12 with those of Barenblatt's model reveals the sensitivity of the material's response to the sign of the initial curvature $\theta''(0)$ of the cohesive energy. Indeed, while an initially concave energy determines the opening of a single crack, an initially convex energy leads to the formation of microstructure.

Among the initially concave energies, of special interest are the *bi-modal* [54] and the *periodic* [52] energies. A bi-modal energy is convex in an interval away from the origin, and concave everywhere else. An example is shown in Fig. 13(a), where the convex part of the curve is denoted by HK. To the convex part corresponds the ascending branch of the bridging force-crack opening curve in Fig. 13(b).

The force-elongation response curve in Fig. 13(c) is constructed with the same procedure used for Barenblatt's energy in Fig. 5. A major difference is that here, in addition to a single jump with amplitude not in the convex part of the curve, a stable equilibrium configuration

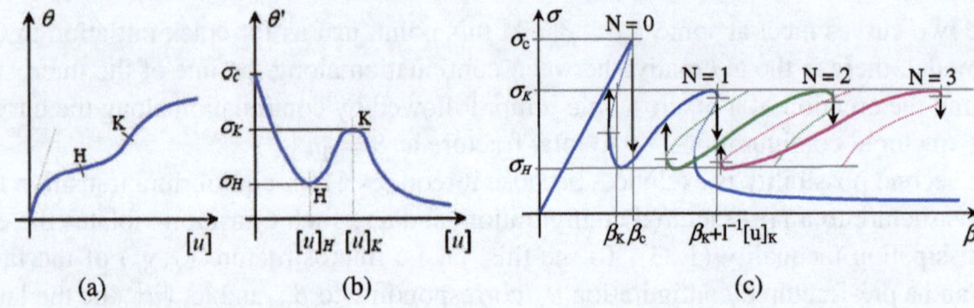

Fig. 13 A bi-modal cohesive energy (**a**), its derivative (**b**), and the force-elongation response curve for small-size bars (**c**). In (**c**), *bold* and *light lines* denote stable and unstable equilibrium configurations, and *downward* and *upward arrows* denote non-equilibrium transitions at loading and unloading, respectively

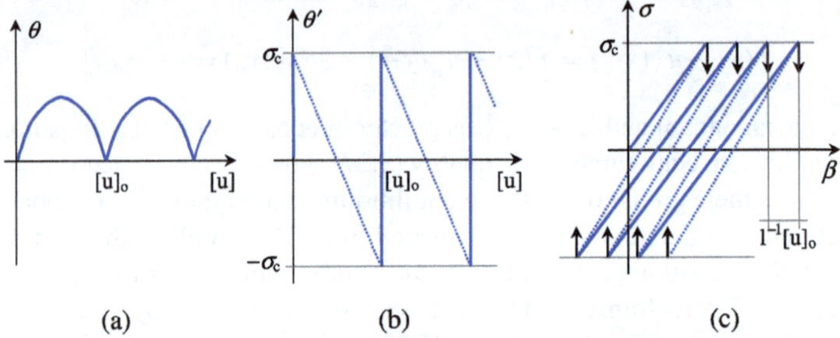

Fig. 14 A periodic cohesive energy (**a**), its derivative (**b**), and the force-elongation response curve (**c**). In (**c**), *solid* and *dashed lines* denote stable and unstable equilibrium configurations, and *downward* and *upward arrows* denote nonequilibrium transitions at loading and unloading, respectively

may have an arbitrary number of jumps with amplitude in the convex part of the curve. The stability analysis is more complicated than in the case of a concave energy, see [63, Sect. 3.4]. In Fig. 13(c), the stable and unstable parts of the equilibrium curves are traced in bold and in light, respectively.

For growing β, after a peak at $\beta = \beta_c$, the response curve oscillates between the local minimum σ_H and the local maximum σ_K of the force-crack opening curve. At the N-th oscillation the maximum σ_K is attained at

$$\beta_N = (w')^{-1}(\sigma_K) + Nl^{-1}[\![u]\!]_K,$$

and at each such β_N a new crack opens. Then the number N of the oscillations coincides with the number of the opened cracks. Each crack opening increases the level of damage, as shown by the slope of the response curve, which decreases with increasing N.

An oscillatory response is also obtained from *periodic* cohesive energies [52]. This class of energies properly describes the microslips occurring in single crystals containing a large number of parallel slip bands [42]. In the energy represented in Fig. 14(a), the presence of a corner point for θ at each multiple of the period $[\![u]\!]_o$ imposes to the bridging force σ the double-side bound

$$-\sigma_c \leq \sigma \leq \sigma_c.$$

It determines the *elastic range* $[-\sigma_c, \sigma_c]$ shown in Fig. 14(b). The force-elongation response curve in Fig. 14(c) consists of the family of straight segments

$$\sigma = w''(\beta_c)(\beta - \beta_N), \quad \beta_N = Nl^{-1}[\![u]\!]_o, \quad |\sigma| \leq \sigma_c, \quad N = 0, 1, 2, \ldots$$

with β_c as in (2.10). The inelastic deformation, instead of being due to the formation of microstructure as in the case of convex-concave energies, is due to non-equilibrium transitions of small, but finite, amplitude $l^{-1}[\![u]\!]_o$. This is the way in which the present model represents plastic slip. At each transition, the energy $\sigma_c[\![u]\!]_o$ is dissipated. At unloading, when the force reaches the lower limit $-\sigma_c$, the same amount of dissipation is produced under a compressive force. This is an efficient way for describing elastic-plastic hysteresis [42].

A richer response is obtained from *wiggly energies* of the form

$$\theta = \theta_m + \theta_p,$$

with θ_m monotonic increasing and θ_p periodic [52]. With such energies, the perfectly plastic response of Fig. 14(c) can be replaced by a work-hardening or a strain-softening response, or by any combination of the two.

3.7 References and Comments

Microstructures defined as limits of uniformly convergent sequences of ordinary configurations were introduced in [57] under the name of *structured deformations*. The subject was developed in a series of papers, see [58, 63, 111] and in the lecture notes [59, 60]. In particular, a three-dimensional version of the expression (3.12) for the density of microstructure was given in [58]. Below, we briefly comment on the modeling of microstructures by means of non-concave cohesive energies.

1. The regularity of the macroscopic deformation f assumed in Sect. 3.2 can be released. An extension to functions f with a finite or countable number of jumps is immediate [53]. A weaker regularity makes possible to describe some peculiar microstructures. For example, from the function f_2 in Fig. 11(b), in the limit for $(b-a) \to 0$ a *concentrated microstructure* is obtained [60, p. 173], with the whole microstructure concentrated at a single point. Another interesting type of microstructure is obtained by taking as f the sum of g_c plus the Cantor function [34, 57]. In this case the microstructure localizes at the Cantor set, which is a fractal set. The introduction of fractal dimension seems to be very promising for the solution of problems in fracture mechanics [36, 100].

2. Since the 1970's, the study of microstructures gained growing importance in continuum mechanics. Examples are the *fine mixtures* of solid phases in multi-phase materials [8], the director theories for liquid crystals [124, 133], and the models related with strain-softening in plasticity, such as localization of plastic deformation [120], formation of shear bands [16, 119], and necking [127]. A large list of microstructures of interest in mechanics and a detailed analysis of their mechanical modeling can be found in the book [30].

3. The *wiggly energies* proposed in [52] follow an idea of James [87], originated from an earlier suggestion of Ericksen. In their original formulation, these energies were addressed to reproduce, in a purely non-linear elastic context, some aspects of the formation of microstructure in the phase changes of metals and alloys [1, 129].

4. From the mathematical viewpoint, fine mixtures of solid phases are represented by Young measures [46, 112], and the formation of microstructure is described by the relaxation of nonconvex functionals [20, 21, 27]. A variational formulation of softening phenomena is given in [28], and a more general procedure for deriving macroscopic models via

relaxation is developed in [105]. A recent overview of the variational modeling of inelastic microstructures may be found in [79].

5. For *bi-modal energies*, the response curve in Fig. 13(b) is relative to the case of small l. In general, the shape of the response curve strongly depends on l and on the values of limit forces σ_H and σ_K. For example, for $\sigma_H = 0$ and θ' piecewise quadratic, it is possible to recover the simplified response curves usually adopted in damage theories, in which each ascending branch lies on a straight line from the origin [54, 62]. With bi-modal energies it is also possible to describe the *multiple cracking* of concrete [125] and of fiber-reinforced cementitious composites [38].

6. The case of *periodic energies* is singular, since it violates the assumption of monotonicity of θ made in Sect. 2.1. This violation allows a crack to open both in tension and in compression, and provides a more complete reproduction of plastic response.

4 Lecture 4. Irreversibility

4.1 Dissipative Character of the Cohesive Energy

In some of the force-elongation relations met in the preceding Lectures, for each load β there is a unique stable equilibrium configuration. Consequently, the response curve is described by a single-valued function $\sigma = \sigma(\beta)$. In this case, when the load β decreases after reaching a maximum at $\beta > \beta_c$, the model's preview is that the curve be traversed backwards. This is not confirmed by experiments: for most materials, the experimental unloading curve is parallel to the loading curve observed in the unfractured regime $\beta < \beta_c$.

There are also cases in which in a certain range of β there are two or more response curves, and the transition from one to the other is a non-equilibrium transition. For example, in case L of Fig. 9(b), for $\beta \in (\beta_p, \beta_c)$ there are two stable equilibrium curves, and for increasing β a non-equilibrium transition from the curve $N = 0$ to the curve $N = 1$ is required at $\beta = \beta_c$. At unloading, the curve $N = 1$ is followed up to $\beta = \beta_p$, where the model's preview is a reversed non-equilibrium transition to the curve $N = 0$. As shown by the arrow in the figure, this transition requires an upward jump of the axial force. An upward jump has never been observed in the experiments.

Also not completely satisfactory is the reversibility of the pre-fracture set assumed in Lecture 2. That is, of complete healing of a pre-fracture point, when the crack opening reduces to zero. Thus, though the cohesive energy model effectively reproduces the response at loading, there are reasons for re-considering its predictions for unloading.

A sharp improvement comes from the assumption that the cohesive energy is dissipative, that is, that the power spent for crack opening is non-negative. In the Barenblatt-Dugdale model, by differentiation of (2.4), in any deformation process $t \mapsto u_t$ the total power is

$$\dot{E}_t(u_t) = \frac{d}{dt} E(u_t) = \int_0^l w'\big(u_t'(x)\big)\dot{u}_t' dx + \sum_i \theta'\big([\![u_t]\!](x_i)\big)[\![\dot{u}_t]\!](x_i).$$

The integral is the power spent in the unfractured part of the bar, and the sum is the power spent at the pre-fractured points. We now make the assumption that the first is stored as elastic strain energy, while the second is totally dissipated. This assumption is expressed by the inequality

$$\theta'\big([\![u]\!](x)\big)[\![\dot{u}]\!](x) \geq 0, \tag{4.1}$$

to be satisfied at every pre-fractured point x of the bar and in every deformation process. Because θ' is supposed to be positive, this condition reduces to the *dissipation inequality*

$$[\![\dot{u}]\!](x) \geq 0. \tag{4.2}$$

On the contrary, at a totally fractured point the cohesive energy keeps the constant value θ_r in any deformation process. Then condition (4.1) is identically satisfied as an equality, and the dissipation inequality (4.2) does not hold.

Condition (4.2) states that at a pre-fractured point the jump amplitude cannot decrease. By consequence, not only a fracture, but also a pre-fracture cannot heal. Thus, the *irreversibility of fracture*, assumed in Griffith's model and abandoned for pre-fractures in Barenblatt's model, is now re-obtained as a consequence of the dissipation hypothesis.

The dissipation hypothesis strongly modifies the bar's response. Below, we show that a first consequence of this assumption is a huge increase of the number of the pre-fractured equilibrium configurations.

4.2 Pre-Fractured Equilibrium Configurations

Like in the non-dissipative model, in the dissipative model a *pre-fractured configuration* is a configuration in which all points are either unfractured or pre-fractured points. A pre-fractured configuration is represented by a function with the same regularity assumed in Lecture 2 for the non-dissipative model, with jumps $[\![u]\!](x_i) < [\![u]\!]_r$ at a finite set $\{x_i\}$ of jump points, and the partially fractured points coincide with the jump points of u. In the following our analysis will be restricted to pre-fractured configurations.

For a pre-fractured configuration u, the energy has the expression (2.4), and u is an equilibrium configuration if the first variation (2.6) is non-negative. What changes in the dissipative model is the class of the admissible perturbations. Indeed, in place of the non-interpenetration condition (1.23), the dissipation inequality (4.2) now imposes the stronger requirement

$$[\![\eta]\!](x) \geq 0 \quad \forall x \in (0, l). \tag{4.3}$$

As a consequence, the first variation is now evaluated over a more restricted set of perturbations. Therefore, all equilibrium configurations for the non-dissipative model are equilibrium configurations for the dissipative model. In addition, new equilibrium configurations become possible.

And in fact, taking a perturbation η with a single jump at a jump point x_i of u, the non-negativeness of the first variation yields the inequality $w'(u') \leq \theta'([\![u]\!](x_i))$ instead of Eq. (2.8). Combined with the equilibrium condition (2.7), this inequality leads to the definition of an equilibrium configuration as a configuration with constant u', such that

$$w'(u') \leq \min\{\theta'(0), \theta'([\![u]\!](x_i))\}. \tag{4.4}$$

Consider, for example, a concave cohesive energy. For it, as in Lecture 2, the configuration space can be reduced to the plane $(u', [\![u]\!])$. The equilibrium region for an energy of the type S is shown in Fig. 15. It consists of the region below the equilibrium curve (2.15), plus the half-line $\{u' = 0, [\![u]\!] \geq 0\}$ of the totally fractured configurations.

Thus, while in the non-dissipative model the equilibrium configurations form equilibrium curves, in the dissipative model they occupy a two-dimensional region. The equilibrium curves of the non-dissipative model belong to the boundary of this region, since for them inequality (4.4) is satisfied as an equality.

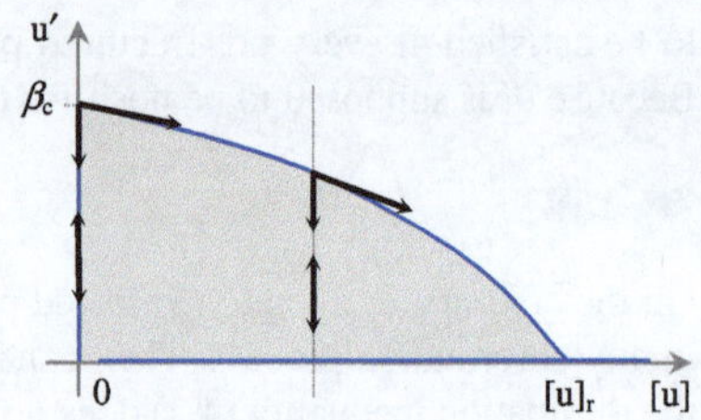

Fig. 15 The dissipative model. The equilibrium region in the $(u', [\![u]\!])$ plane for a concave cohesive energy of the type S. *The arrows* show the directions of the quasi-static evolutions, as determined in Sect. 4.4

4.3 Stability

Like in Griffith's model and in the non-dissipative Barenblatt model, a *stable* configuration u is a local minimizer for the energy with respect to the distance (1.30), in the class of all configurations compatible with u and corresponding to the same β. The restriction (4.3) on the class of admissible perturbations leads to a more restricted notion of compatibility. Indeed, according to (4.3), a pre-fractured configuration v is compatible with u if

$$[\![v]\!](x) \geq [\![u]\!](x) \quad \forall x \in (0, l). \tag{4.5}$$

Consequently, a stable configuration for the non-dissipative model is also stable for the dissipative model. In particular, all unfractured equilibrium configurations are stable by Proposition 2.1, and for pre-fractured configurations with $\#u = 1$ and $w'(u') = \theta'([\![u]\!])$ the stability condition (2.14) of Proposition 2.3 holds. On the contrary, the instability proof of Proposition 2.2 for configurations with $\#u > 1$ does not hold anymore, since it involves negative jumps of η, now forbidden by the dissipation inequality.

For pre-fractured configurations located at the interior points of the equilibrium region, the following stability result holds.

Proposition 4.1 *Let w and θ be as in Proposition* 2.1. *Then all pre-fractured configurations u which satisfy the strict inequality* (4.4) *are stable.*

Proof By the convexity of w and the boundary condition (1.25),

$$\int_0^l \big(w\big(u' + \eta'(x)\big) - w(u')\big)dx \geq w'(u') \int_0^l \eta'(x)dx = -w'(u') \sum_i [\![\eta]\!]_i.$$

Therefore,

$$\begin{aligned} E(u+\eta) - E(u) &= \int_0^l \big(w\big(u' + \eta'(x)\big) - w(u')\big)dx + \sum_i \big(\theta([\![u]\!]_i + [\![\eta]\!]_i) - \theta([\![u]\!]_i)\big) \\ &\geq \sum_i \big(\big(\theta'([\![u]\!]_i) - w'(u')\big)[\![\eta]\!]_i + o([\![\eta]\!]_i)\big) \\ &\geq \min_i \big(\theta'([\![u]\!]_i) - w'(u')\big) \sum_i [\![\eta]\!]_i + o\Big(\sum_i [\![\eta]\!]_i\Big). \end{aligned}$$

Here $[\![u]\!]_i$ and $[\![\eta]\!]_i$ are the jumps of u and η at the jump points of η, respectively. In the last inequality, the transition from the terms $o(\cdot)$ of the individual jumps $[\![\eta]\!]_i$ to the $o(\cdot)$ of their sum is justified by the fact that all $[\![\eta]\!]_i$ are non-negative by (4.3).

On the right side, $[\![u]\!]_i$ is zero at points at which a new jump is created in the perturbed configuration, and is positive otherwise. In all cases, the number of the $[\![\eta]\!]_i$ is finite, and for each of them the term within parenthesis is positive, because inequality (4.4) is assumed to be strict. Consequently, there is an $\varepsilon > 0$ such that the right-hand side is positive for all perturbations for which the sum of all jump amplitudes $[\![\eta]\!]_i$ is smaller than ε. Because, by (1.25) and (1.30),

$$[\![\eta]\!] + \sum_i [\![\eta]\!]_i = -\int_0^l \eta'(x)dx \le l\|\eta\|,$$

the difference $E(u+\eta) - E(u)$ is positive for all η with $l\|\eta\| < \varepsilon$. That is, u is a local minimizer for E. □

4.4 Quasi-Static Evolutions

For the non-dissipative model, our analysis of the evolution of the bar's response under a varying load has been rather informal. For the dissipative model, we go back to the definitions given in Sect. 1.5. We recall that, in a deformation process, for every t and for every $\tau > 0$ the configuration at time $(t+\tau)$ is accessible from the configuration at time t, and that an equilibrium process is a deformation process made of equilibrium configurations.

We also recall that continuity is referred to the metric (1.30) and that, in an analysis restricted to pre-fractured configurations, a configuration is described by the displacement function and, consequently, a deformation process is a continuous family $t \mapsto u_t$ of displacements.

Accessibility is defined by two conditions: compatibility, and the energetic dissipation inequality (1.33). In the dissipative model, compatibility is defined by inequality (4.5). For the energy inequality, consider an equilibrium process $t \mapsto u_t$. For every t, the axial strain u_t' is constant over the bar. Then, the expression of the total power reduces to

$$\begin{aligned}\dot{E}_t(u_t) &= l w'(u_t')\dot{u}_t' + \sum_i \theta'\big([\![u_t]\!](x_{it})\big)[\![\dot{u}_t]\!](x_{it})\\ &= l\sigma_t\dot{\beta}_t + \sum_i \big(\theta'\big([\![u_t]\!](x_{it})\big) - w'(u_t')\big)[\![\dot{u}_t]\!](x_{it}),\end{aligned}$$

and condition (1.33) is satisfied only if the sum on the right is non-positive. But in this sum the term within parentheses is non-negative by the equilibrium condition (4.4), and $[\![\dot{u}_t]\!](x_{it})$ is non-negative by the dissipation inequality (4.2). Therefore, condition (1.33) is satisfied if and only if

$$\big(\theta'\big([\![u_t]\!](x_{it})\big) - w'(u_t')\big)[\![\dot{u}_t]\!](x_{it}) = 0 \quad \forall x_{it}. \tag{4.6}$$

This tells us that *in an equilibrium process, there is no change of the jump amplitudes at the points at which inequality* (4.4) *is strict.*

A *quasi-static evolution* is an equilibrium process in which all u_t are stable equilibrium configurations. A basic problem is to determine the quasi-static evolution associated with a given load process $t \mapsto \beta_t$. This can be done by incremental energy minimization. Indeed, suppose that the configuration u_t is known. For a sufficiently small $\tau > 0$, the configuration $u_{t+\tau}$ can be determined by minimizing the energy

$$E_{t+\tau}(v) = l w\Big(\beta_{t+\tau} - l^{-1}\sum_j [\![v]\!](x_j)\Big) + \sum_j \theta\big([\![v]\!](x_j)\big), \tag{4.7}$$

in the set of all equilibrium configurations v corresponding to $\beta_{t+\tau}$, which are accessible from u_t.

Let us consider the case of θ concave. On the basis of the results of the previous Lectures, it is expected that a minimizer has at most one jump. Then, by the accessibility condition (4.5), if u_t has a jump of amplitude $[\![u_t]\!]$ at a point x_o, a minimizer $u_{t+\tau}$, if it exists, must have a single jump at the same point, of amplitude $[\![u_{t+\tau}]\!] \geq [\![u_t]\!]$. For configurations which satisfy this condition, the energy (4.7) takes the simpler form

$$E_{t+\tau}(v) = lw\big(\beta_{t+\tau} - l^{-1}[\![v]\!]\big) + \theta([\![v]\!]), \qquad [\![v]\!] \geq [\![u_t]\!]. \tag{4.8}$$

In this way, the energy becomes a function of a single scalar variable, the jump amplitude $[\![v]\!]$. The specification of the position of the jump is omitted, because it is irrelevant for the minimization. An approximate determination of the minimizer is obtained by discretizing the variable t. Select a time step $\tau > 0$, and set

$$[\![v]\!] = [\![u_t]\!] + \tau[\![\dot{u}]\!], \qquad \beta_{t+\tau} = \beta_t + \tau\dot{\beta}_t. \tag{4.9}$$

Then consider the expansion

$$E_{t+\tau}(v) = E_t(u_t) + \tau I_t([\![\dot{u}]\!]) + \frac{1}{2}\tau^2 J_t([\![\dot{u}]\!]) + o(\tau^2), \tag{4.10}$$

where, by (4.8),

$$\begin{aligned} I_t([\![\dot{u}]\!]) &= l\sigma_t\dot{\beta}_t + \big(\theta'([\![u_t]\!]) - \sigma_t\big)[\![\dot{u}]\!], \qquad \sigma_t = w'(u_t'), \\ J_t([\![\dot{u}]\!]) &= lw''(u_t')\dot{\beta}_t^2 - 2w''(u_t')\dot{\beta}_t[\![\dot{u}]\!] + \big(l^{-1}w''(u_t') + \theta''([\![u_t]\!])\big)[\![\dot{u}]\!]^2. \end{aligned} \tag{4.11}$$

The minimizer $[\![\dot{u}_t]\!]$ of the function $I_t + \frac{1}{2}\tau J_t$ determines the continuation $\dot{u}_t$ of the quasi-static evolution at the time t. A first-order approximation is obtained by minimizing the function $I_t(\cdot)$ under the condition

$$[\![\dot{u}]\!] \geq 0. \tag{4.12}$$

This function is the sum of the term $l\sigma_t\dot{\beta}_t$, which is known, plus the product of $[\![\dot{u}]\!]$ by a term which is non-negative by the equilibrium condition (4.4)

$$\theta'([\![u_t]\!]) - \sigma_t \geq 0. \tag{4.13}$$

The minimum is achieved when this product is zero

$$\big(\theta'([\![u_t]\!]) - \sigma_t\big)[\![\dot{u}]\!] = 0. \tag{4.14}$$

This is exactly the accessibility condition (4.6), and (4.12), (4.13) and (4.14) are the Kuhn-Tucker conditions associated with the minimum problem for I_t.

The complementarity condition (4.14) states that $[\![\dot{u}_t]\!]$ is zero if u_t is an interior point of the equilibrium region. If u_t is a boundary point, then $(\theta'([\![u_t]\!]) - \sigma_t)$ is zero, and the Kuhn-Tucker conditions are satisfied by any non-negative $[\![\dot{u}]\!]$. In this case, to determine $[\![\dot{u}_t]\!]$ it is necessary to minimize the second-order term $J_t(\cdot)$ of the expansion (4.10), which is a quadratic function of $[\![\dot{u}]\!]$. If the strict stability condition (2.11) holds, the quadratic part

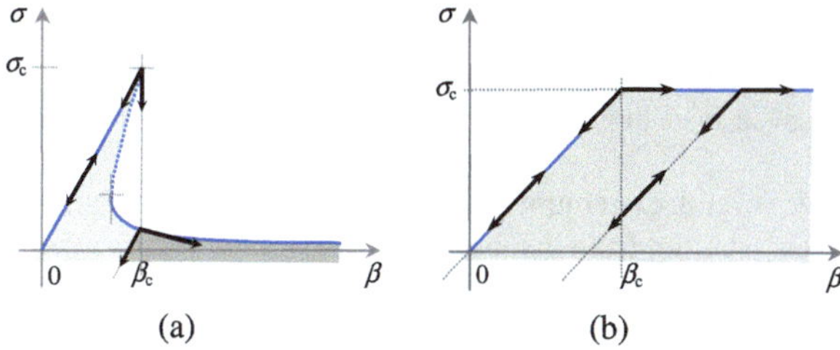

Fig. 16 The dissipative model. Representation of the force-elongation response for a concave cohesive energy and a large-size bar (**a**), and for a convex-concave cohesive energy (**b**). *The gray areas* denote the equilibrium configurations. In (**a**), *the light-gray area* denotes equilibrium configurations not accessible from the origin. *The arrows* show the directions of the quasi-static evolutions at loading and at unloading

of $J_t(\cdot)$ is positive. In the presence of the constraint $[\![\dot{u}]\!] \geq 0$, the minimum is achieved at

$$[\![\dot{u}_t]\!] = \begin{cases} 0 & \text{if } \dot{\beta}_t \leq 0, \\ \frac{w''(u_t')}{\theta''([\![u_t]\!]) + l^{-1} w''(u_t')} \dot{\beta}_t & \text{if } \dot{\beta}_t > 0. \end{cases} \tag{4.15}$$

Consequently,

$$\dot{\sigma}_t = w''(u_t')\big(\dot{\beta}_t - l^{-1}[\![\dot{u}_t]\!]\big) = \begin{cases} w''(u_t')\dot{\beta}_t & \text{if } \dot{\beta}_t \leq 0, \\ \frac{\theta''([\![u_t]\!]) w''(u_t')}{\theta''([\![u_t]\!]) + l^{-1} w''(u_t')} \dot{\beta}_t & \text{if } \dot{\beta}_t > 0. \end{cases} \tag{4.16}$$

This provides the slope $\dot{\sigma}_t/\dot{\beta}_t$ of the response curve at the boundary of the equilibrium region. For $\dot{\beta}_t \leq 0$, the slope $w''(u_t')$ is the same occurring at the interior points. This is the property of *elastic unloading* observed in the response of elastic-plastic materials. For $\dot{\beta}_t > 0$, the slope provided by (4.16) is the slope necessary to continue the evolution remaining on the boundary of the equilibrium region. Indeed, this slope is obtained by differentiation of the equations

$$\sigma_t = w'\big(\beta_t - l^{-1}[\![u_t]\!]\big) = \theta'([\![u_t]\!]),$$

and the slope given by (4.16) follows after elimination of $[\![\dot{u}_t]\!]$.

In Fig. 16(a), the equilibrium region for a concave cohesive energy and a large-size bar is represented in the (σ, β) plane, and the directions of the quasi-static evolution at loading and at unloading are represented by arrows. Due to the non-equilibrium transition at $\beta = \beta_c$, the configurations in the light-gray region are not accessible from the origin. In particular, the portion of the equilibrium curve $N = 1$ with $\beta < \beta_c$ is inaccessible from the origin. This rules out the unrealistic jump at unloading predicted by the non-dissipative model.

For a convex-concave energy, the directions of the quasi-static evolution are shown in Fig. 16(b). They exactly reproduce the response of an elastic-perfectly plastic material.

4.5 References and Comments

The irreversibility assumption completely changes the nature of the quasi-static evolutions. In the non-dissipative case, the equilibrium configurations lie on equlibrium curves, and in a given load process these curves are traversed back and forth, save for the non-equilibrium transitions which arise in some special situations. On the contrary, in the dissipative case the equilibrium configurations form a region in the configuration space. Then there is an infinity of possible equilibrium processes, and to determine the direction of the quasi-static evolution under a given load increment it is necessary to minimize the incremental energy (4.10).

In the literature, several extremum principles involving strain energy plus dissipation have been formulated, starting from the *principle of maximum plastic work* in plasticity [80]. References on the subsequent developments can be found in the paper [78]. In the same paper, two types of extremum principles are considered, the *principle of maximum dissipation* and the *minimum principle for the dissipation potential*, and it is shown by examples that not always do they provide the same evolution equations.

The principle assumed in this Lecture is of the second type, though the name seems inappropriate because what is minimized is not the dissipation potential but its sum with the strain energy. This principle has the advantage of relying on definitions of equilibrium, stability, and quasi-static evolution, which can be easily extended to microstructures.

The incremental procedure adopted here is due to Fedelich and Ehrlacher [68], who were the first to propose higher-order minimization to determine the part of the evolution left arbitrary by the first-order minimization. This technique was applied in [63] and, for finite plasticity, in [104]. The enlargement of the equilibrium set due to the introduction of the dissipation inequality was pointed out in [63], and the connection between dissipation and elastic unloading was established in [51].

The assumption that the cohesive energy is totally dissipative may look too strong. In fact, it is appropriate to describe a plastic response, in which the permanent deformation left after total unloading from a pre-fractured configuration is equal to the sum of the jump amplitudes. On the contrary, in most of damage theories there is no permanent deformation. That is, the response curve at unloading goes back to the origin. This type of response can be captured by assuming that only a part of the cohesive energy is dissipated, see, e.g., [61]. Alternative descriptions are provided by the diffuse energy models described in the next Lectures.

5 Lecture 5. Diffuse Fracture: The Local Model

5.1 Bulk Cohesive Energies

In the regularized approach to fracture discussed in the preceding Lectures, the elastic energy has a density w per unit length, and the cohesive energy θ is defined on isolated points. In a three-dimensional context, the same energies have densities per unit volume and per unit area, respectively.

The dimensional difference between w and θ is motivated by the observation that fracture usually occurs along surfaces. But this motivation is not completely satisfactory, since there are cases in which the opening of a crack is preceded by a spread of inelastic deformations over a volume portion of the body, the *process zone*. This produces a weakening of the material, usually described by damage or plasticity theories, which strongly influences the fracture mode. Indeed, with increasing load the process zone can either localize on a fracture surface and eventually lead to catastrophic fracture, or it can spread over large portions of the body, producing a generalized weakening without any catastrophic event.

The approach presented here describes a model for diffuse fracture, based on the theory developed in the first part of the paper [56]. In a one-dimensional context, at every point x of the bar the axial strain u' is assumed to be the sum of an elastic and an inelastic part

$$u'(x) = \epsilon(x) + \gamma(x). \tag{5.1}$$

The energy is also assumed to be the sum of two parts, an elastic energy depending on ϵ, and a cohesive energy depending on γ

$$E(\varepsilon,\gamma)=\int_0^l w\big(\epsilon(x)\big)dx+\int_0^l \theta\big(\gamma(x)\big)dx. \tag{5.2}$$

Each part has a volume density. That is, the inelastic part of the deformation spreads over regions of finite length, instead of concentrating at a finite set of points as assumed in the preceding models. The extent and the evolution of the inelastic zone under varying load depend on the form of the function θ. In the following Sections we show that different choices of θ may lead both to a uniform distribution of the inelastic deformation and to localization at a singular surface.

For the functions w and θ we keep the assumptions (1.2), (2.1) and the regularity assumed in Sects. 1.1 and 2.1, respectively. Moreover, we assume that θ is strictly increasing and that $\theta_r=+\infty$.

The functions ϵ and γ in the decomposition (5.1) are supposed to be continuous in $[0,l]$. Then (5.1) can be integrated over $(0,l)$, and the boundary condition (1.4) can be given the form

$$\bar{\epsilon}+\bar{\gamma}=\beta, \tag{5.3}$$

with

$$\bar{\epsilon}=\frac{1}{l}\int_0^l \epsilon(x)dx,\qquad \bar{\gamma}=\frac{1}{l}\int_0^l \gamma(x)dx.$$

A pair (ϵ,γ) is a *configuration* of the bar, and for every β Eq. (5.3) defines the configurations corresponding to β.

The inelastic deformation is assumed to be dissipative, that is, the cohesive power

$$\theta'\big(\gamma_t(x)\big)\dot{\gamma}_t(x)$$

is supposed to be non-negative in all deformation processes $t\mapsto(\epsilon_t,\gamma_t)$, at all points x of the bar. By the assumed positiveness of θ', this assumption reduces to the dissipation inequality

$$\dot{\gamma}_t(x)\geq 0, \tag{5.4}$$

to be satisfied at all x and for all t.

5.2 Equilibrium

In the diffuse energy model, the equilibrium configurations are identified with the pairs (ϵ,γ) for which the first variation of the energy

$$\begin{aligned}\delta E(\epsilon,\gamma,\delta\epsilon,\delta\gamma)&=\lim_{\lambda\to 0^+}\frac{1}{\lambda}\big(E(\epsilon+\lambda\delta\epsilon,\gamma+\lambda\delta\gamma)-E(\epsilon,\gamma)\big)\\&=\int_0^l\big(w'\big(\epsilon(x)\big)\delta\epsilon(x)+\theta'\big(\gamma(x)\big)\delta\gamma(x)\big)dx\end{aligned}$$

is non-negative for all perturbations $(\delta\epsilon,\delta\gamma)$ which preserve the length of the bar

$$\delta\bar{\epsilon}+\delta\bar{\gamma}=\frac{1}{l}\int_0^l \delta\epsilon(x)dx+\frac{1}{l}\int_0^l \delta\gamma(x)dx=0, \tag{5.5}$$

and satisfy the compatibility condition

$$\delta\gamma(x) \geq 0 \quad \forall x \in (0,l). \tag{5.6}$$

This condition is the counterpart of condition (4.3) for cohesive energies concentrated at a finite set of points. Note that, because of this condition, the first variation is defined as a limit on the positive λ only, and not on all λ.

For perturbations with $\delta\gamma = 0$, the non-negativeness of the first variation requires that

$$\int_0^l w'\big(\epsilon(x)\big)\delta\epsilon(x)dx \geq 0$$

for all $\delta\varepsilon$ such that

$$\int_0^l \delta\epsilon(x)dx = 0.$$

Then the first variation is non-negative only if the elastic deformation, and therefore the axial force, is constant over the bar

$$\epsilon(x) = \bar{\epsilon}, \qquad \sigma = w'(\bar{\epsilon}). \tag{5.7}$$

By the boundary condition (5.5), the first variation reduces to

$$\delta E(\epsilon,\gamma,\delta\epsilon,\delta\gamma) = \int_0^l \big(\theta'\big(\gamma(x)\big) - \sigma\big)\delta\gamma(x)dx, \tag{5.8}$$

and from inequality (5.6) the condition

$$\sigma \leq \theta'\big(\gamma(x)\big) \quad \forall x \in (0,l) \tag{5.9}$$

follows. Conversely, for every configuration (ϵ,γ) with constant ϵ obeying this condition the first variation is non-negative. Therefore, (ϵ,γ) is an equilibrium configuration if and only if ϵ is constant and σ satisfies inequality (5.9).

5.3 Stability

We say that an equilibrium configuration (ϵ,γ) is *stable* if it is a local minimizer for the energy with respect to the distance induced by the norm

$$\|(\epsilon,\gamma)\| = \sup_{x\in(0,l)} |\epsilon(x)| + \sup_{x\in(0,l)} |\gamma(x)|, \tag{5.10}$$

in the class of all perturbations $(\delta\epsilon,\delta\gamma)$ which satisfy conditions (5.5) and (5.6).

A consequence of the assumed convexity of w is that, by Jensen's inequality (1.13),

$$\int_0^l w\big(\epsilon(x)\big)dx \geq l w(\bar{\epsilon}),$$

and, consequently,

$$E(\epsilon,\gamma) \geq E(\bar{\epsilon},\gamma).$$

Therefore, the search for minimizers can be restricted to the configurations with constant ϵ which correspond to the given β. For them, the energy takes the simpler form

$$E_\beta(\gamma) = lw(\beta - \bar{\gamma}) + \int_0^l \theta\big(\gamma(x)\big)dx. \tag{5.11}$$

By (5.6) and (5.9), the first variation

$$\delta E_\beta(\gamma, \delta\gamma) = \int_0^l \big(\theta'\big(\gamma(x)\big) - w'(\beta - \bar{\gamma})\big)\delta\gamma(x)dx \tag{5.12}$$

is non-negative for all equilibrium configurations. In particular, it is zero for all perturbations with support in the *inelastic zone*

$$\mathcal{J}_\beta(\gamma) = \big\{x \in (0, l) \mid w'(\beta - \bar{\gamma}) = \theta'\big(\gamma(x)\big)\big\}. \tag{5.13}$$

For such perturbations, the second variation

$$\delta^2 E_\beta(\gamma, \delta\gamma) = \int_0^l \big(w''(\beta - \bar{\gamma})\delta\bar{\gamma}^2 + \theta''\big(\gamma(x)\big)\delta\gamma^2(x)\big)dx \tag{5.14}$$

must be non-negative, and for this it is necessary that

$$\theta''\big(\gamma(x)\big) \geq 0 \quad \forall x \in \mathcal{J}_\beta(\gamma). \tag{5.15}$$

Indeed, if $\theta''(\gamma(x)) < -c < 0$ on an interval $\mathcal{I} \subset \mathcal{J}_\beta(\gamma)$, by concentrating the perturbation on a subinterval $\mathcal{I}_\varepsilon$ of $\mathcal{I}$ of length ε

$$\delta\gamma(x) = \begin{cases} \mu & \text{if } x \in \mathcal{I}_\varepsilon, \\ 0 & \text{otherwise,} \end{cases}$$

one has $\delta\bar{\gamma} = \mu\varepsilon/l$ and, therefore,

$$\delta^2 E_\beta(\gamma, \delta\gamma) < \mu^2\big(l^{-1}w''(\beta - \bar{\gamma})\varepsilon^2 - c\varepsilon\big).$$

The right-hand side is negative for sufficiently small ε. Therefore, condition (5.15) is necessary for a local minimum. This condition is satisfied if θ is convex. In this case, by Jensen's inequality,

$$\int_0^l \theta\big(\gamma(x)\big)dx \geq l\theta(\bar{\gamma}), \tag{5.16}$$

and, therefore,

$$E_\beta(\gamma) \geq E_\beta(\bar{\gamma}).$$

If θ is strictly convex, the above inequalities are strict. Then, if θ is strictly convex, *every stable equilibrium configuration is homogeneous*

$$\gamma(x) = \bar{\gamma} \quad \forall x \in (0, l).$$

In this case, the stability condition (5.15) is optimal. Indeed, in Sect. 2.3 of [56] it has been proved that the non-negativeness of the first variation plus

$$\theta''\big(\gamma(x)\big) > 0 \quad \forall x \in (0, l) \tag{5.17}$$

is a sufficient condition for a local minimum.

If θ is convex but not strictly convex, there may be non-homogeneous stable configurations. For every such configuration, the homogeneous configuration $\gamma(x)=\bar{\gamma}$ is a stable configuration with the same energy, and corresponding to the same β.

5.4 Quasi-Static Evolutions

According to the definitions given in Sect. 1.5, a *deformation process* is a continuous path in the configuration space, in which for every $\tau>0$ the configuration at $t+\tau$ is accessible from the configuration at t. An *equilibrium process* is a deformation process made of equilibrium configurations, and a *quasi-static evolution* is an equilibrium process made of stable equilibrium configurations. For the diffuse fracture model we keep the same definitions, with appropriate changes in the involved objects.

In the present model, a configuration is a pair (ϵ,γ), and a deformation process is a continuous family $t\mapsto(\epsilon_t,\gamma_t)$, with continuity referred to the metric induced by the norm (5.10). Accessibility is again defined as compatibility plus the energetic dissipation inequality. Compatibility is expressed by condition (5.6), which implies

$$\gamma_{t+\tau}(x)\geq\gamma_t(x)\quad\forall x\in(0,l),\ \forall\tau>0. \tag{5.18}$$

Moreover, in the energetic dissipation inequality (1.33), $\dot{E}_t$ is given by

$$lw'(\bar{\epsilon}_t)\dot{\bar{\epsilon}}_t+\int_0^l\theta'\big(\gamma_t(x)\big)\dot{\gamma}_t(x)dx=l\sigma_t\dot{\beta}_t+\int_0^l\big(\theta'\big(\gamma_t(x)\big)-w'(\bar{\epsilon}_t)\big)\dot{\gamma}_t(x)dx,$$

as follows from differentiation of (5.2) and from the relations $\sigma_t=w'(\bar{\epsilon}_t)$ and $\dot{\bar{\epsilon}}_t=\dot{\beta}_t-\dot{\bar{\gamma}}_t$. Therefore, inequality (1.33) is satisfied if and only if

$$\int_0^l\big(\theta'\big(\gamma_t(x)\big)-w'(\bar{\epsilon}_t)\big)\dot{\gamma}_t(x)dx\leq 0 \tag{5.19}$$

for all t. For an equilibrium configuration, the integrand function is non-negative. Therefore, in an equilibrium process the energy inequality is satisfied if and only if

$$\big(\theta'\big(\gamma_t(x)\big)-w'(\bar{\epsilon}_t)\big)\dot{\gamma}_t(x)=0\quad\forall x\in(0,l). \tag{5.20}$$

Let us consider the case of θ strictly convex, for which every stable configuration is homogeneous. In this case, a quasi-static evolution is made of homogeneous configurations. In a quasi-static evolution corresponding to a given load process $t\mapsto\beta_t$, each γ_t minimizes the energy

$$E_t(\gamma)=l\big(w(\beta_t-\gamma)+\theta(\gamma)\big) \tag{5.21}$$

among all constant γ which satisfy condition (5.18), that is, such that $\gamma\geq\gamma_t$.

Like in Sect. 4, to determine a quasi-static evolution corresponding to a given load process and starting from a given initial configuration, we proceed by incremental energy minimization. We assume that γ_t is known, and we wish to determine $\gamma_{t+\tau}$ for a conveniently small $\tau>0$. For this, we set

$$\beta_{t+\tau}=\beta_t+\tau\dot{\beta}_t,\qquad\gamma_{t+\tau}=\gamma_t+\tau\dot{\gamma}_t, \tag{5.22}$$

and, for any perturbed configuration γ,

$$\gamma=\gamma_{t+\tau}+\delta\gamma=\gamma_t+\tau(\dot{\gamma}_t+\delta\dot{\gamma})=\gamma_t+\tau\dot{\gamma}. \tag{5.23}$$

Then we consider the expansion

$$E_{t+\tau}(\gamma) = E_t(\gamma_t) + \tau I_t(\dot{\gamma}) + \frac{1}{2}\tau^2 J_t(\dot{\gamma}) + o(\tau^2). \tag{5.24}$$

Because both γ and γ_t are constant, $\dot{\gamma}$ is constant as well. Therefore,

$$\begin{aligned} I_t(\dot{\gamma}) &= l\big(\sigma_t \dot{\beta}_t + (\theta'(\gamma_t) - \sigma_t)\dot{\gamma}\big), \qquad \sigma_t = w'(\beta_t - \gamma_t) = w'(\epsilon_t), \\ J_t(\dot{\gamma}) &= l\big(w''(\epsilon_t)\dot{\beta}_t^2 - 2w''(\epsilon_t)\dot{\beta}_t\dot{\gamma} + (w''(\epsilon_t) + \theta''(\gamma_t))\dot{\gamma}^2\big). \end{aligned} \tag{5.25}$$

Neglecting the term $o(\tau^2)$, the minimization of $E_{t+\tau}$ is reduced to the minimization of a quadratic function of the scalar variable $\dot{\gamma}$ under the constraint $\dot{\gamma} \geq 0$. The minimum is achieved at

$$\dot{\gamma}_t = \max\left\{ \frac{-\theta'(\gamma_t) + \sigma_t + \tau w''(\epsilon_t)\dot{\beta}_t}{\tau(\theta''(\gamma_t) + w''(\epsilon_t))}, 0 \right\}. \tag{5.26}$$

Because in the fraction the denominator is positive, the solution depends on the sign of the numerator. In the elastic regime, $\theta'(\gamma_t) > \sigma_t$, the numerator is negative for sufficiently small τ. The solution is $\dot{\gamma}_t = 0$, and the stress increment is

$$\dot{\sigma}_t = w''(\epsilon_t)\dot{\beta}_t. \tag{5.27}$$

In the inelastic regime, $\theta'(\gamma_t) = \sigma_t$, the solution is

$$\dot{\gamma}_t = \begin{cases} 0 & \text{if } \dot{\beta}_t \leq 0, \\ \frac{w''(\epsilon_t)}{\theta''(\gamma_t) + w''(\epsilon_t)}\dot{\beta}_t & \text{if } \dot{\beta}_t > 0, \end{cases} \tag{5.28}$$

and the stress increment is

$$\dot{\sigma}_t = w''(\epsilon_t)(\dot{\beta}_t - \dot{\gamma}_t) = \begin{cases} w''(\epsilon_t)\dot{\beta}_t & \text{if } \dot{\beta}_t \leq 0, \\ \frac{\theta''(\gamma_t)w''(\epsilon_t)}{\theta''(\gamma_t) + w''(\epsilon_t)}\dot{\beta}_t & \text{if } \dot{\beta}_t > 0. \end{cases} \tag{5.29}$$

The analogy with Eqs. (4.15) and (4.16), is only formal, because in the previous model the inelastic deformation concentrates on a single jump point, while here it smears uniformly over the bar.

Equations (5.27) and (5.29) show that the slope $\dot{\sigma}_t/\dot{\beta}_t$ of the force-elongation response curve is equal to $w''(\bar{\varepsilon}_t)$ in the elastic regime and in the inelastic regime at unloading, $\dot{\beta}_t \leq 0$. In the inelastic regime at loading it varies between zero and $w''(\bar{\varepsilon}_t)$, depending on the current values of w'' and θ''. It is zero in the limit for $\theta''(\gamma_t) \to 0$, and it is the same as in the elastic regime for $\theta''(\gamma_t) \to +\infty$.

The relation (5.28) provides the differential equation

$$\big(\theta''(\gamma_t) + w''(\epsilon_t)\big)\dot{\gamma}_t = w''(\epsilon_t)\dot{\beta}_t, \tag{5.30}$$

which governs the growth of the inelastic deformation in the inelastic regime at loading. This equation holds as long as $\theta''(\gamma_t)$ stays positive. When it becomes zero, the restriction to the constant $\dot{\gamma}$ is no longer possible. Indeed, for $\sigma_t = \theta'(\gamma_t)$ and $\theta''(\gamma_t) = 0$ the functions I_t and J_t reduce to

$$I_t(\dot{\gamma}) = l\sigma_t\dot{\beta}_t, \qquad J_t(\dot{\gamma}) = l w''(\epsilon_t)\big(\dot{\beta}_t - \bar{\dot{\gamma}}^2\big).$$

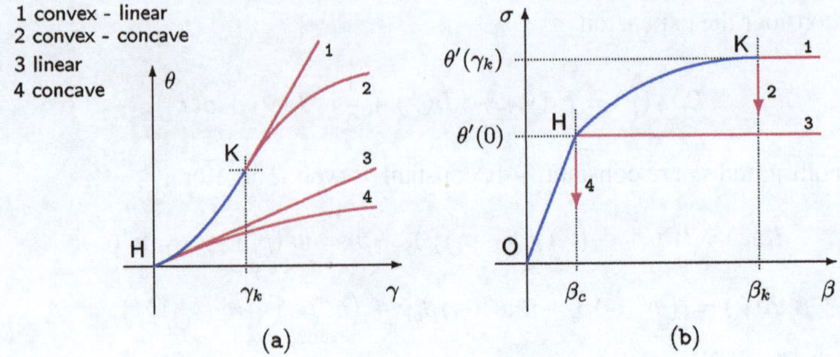

Fig. 17 The local model of diffuse fracture. Some forms of the cohesive energy (**a**), and the corresponding response curves for t (**b**). *The arrows* denote catastrophic rupture

The minimum is achieved at $\bar{\dot{\gamma}}_t = \dot{\beta}_t$, which corresponds to $\dot{\sigma}_t = 0$, that is, to *perfectly plastic response*. In this case the incremental minimization specifies the average $\bar{\dot{\gamma}}_t$, but the punctual values $\dot{\gamma}_t(x)$ are left undetermined. Thus, *in a perfectly plastic response, the evolution from a homogeneous configuration need not be homogeneous.*

When $\theta''(\gamma_t)$ becomes negative, it has been shown in Sect. 5.3 that a negative second variation is obtained from perturbations concentrated on intervals of sufficiently small length. This extreme concentration of inelastic deformations determines the catastrophic rupture of the bar.

Some possible forms of the cohesive energy are shown in Fig. 17(a). The point K marks the transition from convex to linear for the energy (1), and from convex to concave for the energy (2). The energies (3) and (4) are linear and concave, respectively. In the force-elongation response curves of Fig. 17(b), the line OH represents the initial elastic regime $\sigma_t = w'(\beta_t)$, which is the same for all θ. The inelastic regime starts at the point H at which β reaches the value $\beta_c = w'^{-1}(\theta'(0))$. At this point, by (5.29), the slope of the curve switches from $w''(\beta_c)$ to

$$\frac{\theta''(0)w''(\beta_c)}{\theta''(0) + w''(\beta_c)}.$$

The inelastic response at H is work-hardening for the energies (1) and (2), and perfectly plastic for (3). For the concave energy (4) the stability condition $\theta''(\gamma) \geq 0$ is violated, and catastrophic rupture occurs just at the beginning of the inelastic regime. For the energy (1), the response becomes perfectly plastic when β reaches the value $\beta_k = \beta_c + \gamma_k$. At this same value the energy (2) becomes concave, and catastrophic rupture takes place.

5.5 References and Comments

For decades, the research in fracture mechanics was concentrated on the study of the stress singularity at the tip of a crack. This interest was justified by the urgency of understanding the causes of catastrophic failure, and Griffith's paper [76] gave a formidable momentum in this direction. The goal has now been achieved, but at the price of leaving aside a more comprehensive view of fracture including ductile and ductile-brittle collapse mechanisms.

By contrast, the diffuse model describes fracture as the terminal event of a progressive weakening process involving volume portions of the body. A first step in this direction was the inclusion of the work of plastic deformation in the energy balance, see, e.g., the survey [136] on the contribution of G.R. Irwin to fracture mechanics. On the side of mathematical analysis, the already mentioned model of Ambrosio and Tortorelli [6] introduced the representation of the fracture energy by a volume integral as a regularizing approximation. More recent is the interest in diffuse cohesive energies as genuine physical models [56, 71, 115, 134]. A presentation of local and non-local models for damage may be found in [113] and [114], respectively.

A serious limit for the local model is the failure to describe softening. This is due to the excessive severity of the stability condition (5.15). Indeed, combining Eqs. (5.27) and (5.28), at loading one has

$$\dot{\sigma}_t = \theta''(\gamma_t)\dot{\gamma}_t, \tag{5.31}$$

and multiplying by $\dot{\gamma}_t$, for a non-negative $\theta''(\gamma_t)$ one has

$$\dot{\sigma}_t\dot{\gamma}_t = \theta''(\gamma_t)\dot{\gamma}_t^2 \geq 0. \tag{5.32}$$

Then $\dot{\sigma}_t$ must be non-negative when $\dot{\gamma}_t > 0$, and this rules out the possibility of a strain-softening response. Inequality (5.32) is Drucker's *material stability postulate* [64]. Therefore, the local model works only for materials which obey this postulate.

In fact, it comes out that *the local model describes the classical plasticity theory based on Drucker's postulate*. Indeed, this theory is based on a number of assumptions

- *the yield condition*,
- *the hardening rule*,
- *the consistency condition*,
- *the loading-unloading law*,

which translate into mathematical language the phenomenological aspects typical of plastic behavior.

In the present model, all these assumptions are *deduced* as necessary conditions for a minimum in the incremental energy minimization. Indeed, the equilibrium condition (5.9) states that there is a threshold for σ, depending on the punctual values of γ. In the language of plasticity, this is a *yield condition*, $\theta'(\gamma(x))$ is the *yield force*, the difference $f(\sigma,\gamma) = \theta'(\gamma) - \sigma$ is the *yield function*, and $f(\sigma,\gamma) = 0$ is the equation of the *yield surface*. The *hardening rule* is the relation (5.31), which specifies the increment of σ as a function of the current value of γ and of its increment $\dot{\gamma}$. The *consistency condition* is the same relation (5.31), rewritten in the form

$$\dot{f}(\sigma_t,\gamma_t) = 0. \tag{5.33}$$

It states the permanence on the yield surface in the inelastic regime at loading. Finally, the *loading-unloading law* is expressed by the alternative in Eq. (5.29).

As shown by the response curves of Fig. 17(b), within the limits imposed by Drucker's postulate the model's response is quite flexible, and very sensitive to the shape of the function θ. The same response curves show two possible ways for terminating a monotonic loading process: catastrophic fracture in cases (2) and (4), and unlimited perfectly plastic response in cases (1) and (3). A third way, not shown in the figure, is the unlimited work-hardening response provided by a strictly convex θ.

The local model has two weak points: it does not describe the ductile fracture mode, and it exhibits no size effect. These inconveniences are generally attributed to the fact that the model does not include any *internal length* of the material. As shown in the next Lecture, this unfavorable aspect can be eliminated by the introduction of a non-local energy term.

6 Lecture 6. Diffuse Fracture: The Non-local Model

6.1 Basic Assumptions

The non-local model is obtained by adding to the energy (5.2) a term proportional to the square of the derivative of γ

$$E(\epsilon,\gamma)=\int_0^l\Big(w\big(\epsilon(x)\big)+\theta\big(\gamma(x)\big)+\frac{1}{2}\alpha\gamma'^2(x)\Big)dx, \tag{6.1}$$

with α a positive material constant. The non-local character of the model is due to the fact that the integrand function, which is the energy density at a point x, does not depend only on the values of ϵ and γ at x, but also, via the gradient term, on the values taken by γ at the neighboring points. Technically, this is a *weakly non-local model of the gradient type* [15, 121], and the gradient term is a *singular perturbation* [26, 27].

The regularity of the energy densities w and θ is assumed to be the same as in the local model. On the contrary, the new term requires more regularity for the function γ. We assume that γ is continuous, with a continuous derivative, and with a piecewise continuous second derivative. That is, γ'' is supposed to be continuous, except at a finite number of points x_i, at which the left and the right limits exist.

Like in the local model, the inelastic deformation is supposed to be dissipative, that is, at all $x\in(0,l)$ the cohesive power

$$\theta'\big(\gamma_t(x)\big)\dot\gamma_t(x)+\alpha\gamma_t'(x)\dot\gamma_t'(x)$$

is supposed to be non-negative in all processes $t\mapsto\gamma_t$. For this, it is necessary that the product $\theta'(\gamma_t(x))\dot\gamma_t(x)$ be non-negative at all x. Indeed, if it is negative at some x_o, since the derivative $\dot\gamma_t'(x_o)$ is free to take opposite signs in different deformation processes initiating at γ_t, for at least one choice the cohesive power is negative. Thus, keeping the constitutive assumption of θ' positive, the inequality $\dot\gamma_t(x)\geq 0$ is re-obtained as a necessary condition for dissipativity.

6.2 Equilibrium

Like in the local model, an equilibrium configuration is identified with a configuration (ϵ,γ) for which the first variation of the energy, now given by

$$\delta E(\epsilon,\gamma,\delta\epsilon,\delta\gamma)=\int_0^l\big(w'\big(\epsilon(x)\big)\delta\epsilon(x)+\theta'\big(\gamma(x)\big)\delta\gamma(x)+\alpha\gamma'(x)\delta\gamma'(x)\big)dx$$

is non-negative for all perturbations $(\delta\epsilon,\delta\gamma)$ which satisfy the boundary condition (5.5) and the compatibility condition (5.6). Proceeding as in the previous Lecture, for $\delta\gamma=0$ we find that ϵ must be constant over the bar, $\epsilon(x)=\bar\epsilon$. Then, setting $\sigma=w'(\bar\epsilon)$, from the boundary condition (5.5) we get

$$\int_0^l w'\big(\epsilon(x)\big)\delta\epsilon(x)dx=w'(\bar\epsilon)\int_0^l\delta\epsilon(x)dx=-\sigma\int_0^l\delta\gamma(x)d\xi,$$

and, after integrating by parts the gradient term, we find that

$$\delta E(\epsilon,\gamma,\delta\epsilon,\delta\gamma)=\int_0^l\big(\theta'\big(\gamma(x)\big)-\sigma-\alpha\gamma''(x)\big)\delta\gamma(x)dx+\alpha\big[\gamma'(x)\delta\gamma(x)\big]_0^l.$$

Let us define

$$f(\sigma, \gamma(x), \gamma''(x)) = \theta'(\gamma) - \sigma - \alpha\gamma''. \tag{6.2}$$

Then the non-negativeness of the first variation requires the inequality

$$f(\sigma, \gamma(x), \gamma''(x)) \geq 0 \tag{6.3}$$

at all interior points of $(0, l)$, and the conditions

$$\gamma'(l)\delta\gamma(l) \geq 0, \qquad \gamma'(0)\delta\gamma(0) \leq 0, \tag{6.4}$$

at the boundary. The function f is the *yield function* for the non-local model, and inequality (6.3) is the non-local version of the yield condition (5.9). The difference with the local model is that, while the yield force $\theta'(\gamma(x))$ of the local model depends only on the value of γ at x, with the addition of the non-local term $\alpha\gamma''(x)$ the yield force at x also depends on the values of γ at the neighboring points.

At the boundary, conditions (6.4) can be satisfied in two ways. If $\delta\gamma$ is allowed to take arbitrary positive values at the boundary, the conditions

$$\gamma'(l) \geq 0, \qquad \gamma'(0) \leq 0, \tag{6.5}$$

follow. The second possibility is to impose

$$\gamma(l) = \gamma(0) = 0, \tag{6.6}$$

and, consequently,

$$\delta\gamma(l) = \delta\gamma(0) = 0. \tag{6.7}$$

While conditions (6.6) produce solutions with the inelastic deformation concentrating away from the boundary, with conditions (6.5) the inelastic deformation tends to concentrate at the boundary. In laboratory tests this effect is carefully avoided, either by reinforcing the specimen's end sections or by weakening the central part of the bar with the creation of a notch. For this reason we prefer the boundary conditions of the second type, which give the possibility of a comparison with experiments. With this choice, we have that *an equilibrium configuration for the non-local model is a pair $(\bar{\epsilon}, \gamma)$ with constant $\bar{\epsilon}$, and with γ satisfying inequality* (6.3) *at the interior points and conditions* (6.6) *at the boundary.*

6.3 Stability

An equilibrium configuration (β, γ) is stable if it is a local minimizer for the energy with respect to the norm (5.10), in the class of all perturbations which satisfy the compatibility condition (5.6) and the boundary conditions (5.5) and (6.7). Like in the local model, due to the convexity of w, the search for minimizers can be restricted to the configurations with constant ϵ, corresponding to the given β. For such configurations, the energy has the form

$$E_\beta(\gamma) = lw(\beta - \bar{\gamma}) + \int_0^l \left(\theta(\gamma(x)) + \frac{1}{2}\alpha\gamma'^2(x)\right)dx. \tag{6.8}$$

The first variation

$$\delta E_\beta(\gamma, \delta\gamma) = \int_0^l (\theta'(\gamma(x)) - w'(\beta - \bar{\gamma}) - \alpha\gamma''(x))\delta\gamma(x)dx$$

is non-negative for all equilibrium configurations. Indeed, by (6.2), the integrand function is the product $f\delta\gamma$, with f non-negative by the equilibrium condition (6.3) and $\delta\gamma$ non-negative by the compatibility condition (5.6). Then, a necessary condition for a minimum at γ is that the second variation

$$\delta^2 E_\beta(\gamma,\delta\gamma) = \int_0^l \big(w''(\beta-\bar{\gamma})\delta\bar{\gamma}^2 + \theta''(\gamma(x))\delta\gamma^2(x) + \alpha\delta\gamma'^2(x)\big)dx \tag{6.9}$$

be non-negative for all admissible perturbations $\delta\gamma$ for which the first variation is zero. In particular, the first variation is zero for all perturbations with support in the inelastic zone, now defined by

$$\mathcal{J}_\beta(\gamma) = \big\{x \in (0,l) \mid f\big(\sigma,\gamma(x),\gamma''(x)\big) = 0\big\}.$$

Thus, a necessary condition for a minimum is that the second variation be non-negative for all perturbations with support in $\mathcal{J}_\beta(\gamma)$. Let $(a, a+l_j)$ be an interval in this region. Then a necessary condition for a minimum is the non-negativeness of the smallest eigenvalue ρ_1 of the eigenvalue problem

$$lw''(\beta-\bar{\gamma})\delta\bar{\gamma}^2 + \int_a^{a+l_j} \big(\theta''(\gamma(x))\delta\gamma^2(x) + \alpha\delta\gamma'^2(x)\big)dx = \rho\int_a^{a+l_j} \delta\gamma^2(x)dx, \tag{6.10}$$

over all perturbations $\delta\gamma \geq 0$ with support in $(a, a+l_j)$. In [56] it has been proved that ρ_1 is non-negative only if

$$\alpha\lambda_j^2 + \theta_j'' \geq 0, \tag{6.11}$$

where the negative constant θ_j'' is a weighted average of $\theta''(\gamma(x))$ in $(a, a+l_j)$, and λ_j is a solution of the equation

$$\lambda_j^2 = \frac{l_j}{l}\frac{w''(\bar{\epsilon})}{\alpha}\psi_o(\lambda_j l_j), \tag{6.12}$$

with

$$\psi_o(z) = \begin{cases} 1 - \frac{\tan z/2}{z/2} & \text{if } \pi \leq z < 2\pi, \\ \frac{2\pi}{z} & \text{if } z \geq 2\pi. \end{cases} \tag{6.13}$$

The function ψ_o is continuous, decreasing, and has a continuous derivative. From its graph, shown in Fig. 18, one sees that Eq. (6.12) has a unique solution $\lambda_j l_j > \pi$ for every value of the positive constant

$$c = \frac{w''(\bar{\epsilon})}{\alpha l}.$$

Moreover, λ_j is a decreasing function of l_j. Indeed, rewriting Eq. (6.12) in the form

$$\lambda_j^3 = cz\psi_o(z),$$

with $z = \lambda_j l_j$, and observing that the function $z \mapsto z\psi_o(z)$ is non-increasing, we have that λ_j is a non-increasing function of z. Therefore,

$$0 \geq \frac{dz}{d\lambda_j} = \frac{d(\lambda_j l_j)}{d\lambda_j} = l_j + \lambda_j\frac{dl_j}{d\lambda_j},$$

and $dl_j/d\lambda_j < 0$ follows form the positiveness of l_j and λ_j. By consequence, if condition (6.11) is satisfied for the largest interval contained in the inelastic zone, it is satisfied for all smaller intervals.

In [56] it has also been proved that a sufficient condition for a minimum for E_β is that

$$\alpha\lambda_l^2 + \theta''_{\min} > 0, \tag{6.14}$$

with λ_l the solution of (6.12) for $l_j = l$, and $\theta''_{\min}$ the smallest value of $\theta''(\gamma(x))$ in $(0, l)$.

Conditions (6.11) and (6.14) may be very far from each other. Anyway, they show that the presence of the gradient term has a stabilizing effect, since some equilibrium configurations with moderate negative values of θ'' may now be stable. This is essentially what renders possible a description of the softening response within the present model.

6.4 Quasi-Static Evolutions

For the non-local model, a deformation process is a path $t \mapsto (\epsilon_t, \gamma_t)$ in the configuration space, continuous with respect to the norm (5.10), and satisfying the compatibility condition (5.18) and the energetic dissipation inequality (1.33), with

$$\begin{aligned}\dot{E}_t &= l w'(\bar{\epsilon}_t)(\dot{\beta}_t - \dot{\bar{\gamma}}_t) + \int_0^l \big(\theta(\gamma_t(x)) - \alpha\gamma_t''(x)\big)\dot{\gamma}_t(x)dx \\ &= l\sigma_t\dot{\beta}_t + \int_0^l f_t(x)\dot{\gamma}_t(x)dx.\end{aligned}$$

The last equality is obtained by integration by parts, using the boundary conditions (6.6) and setting

$$\sigma_t = w'(\bar{\epsilon}_t), \qquad \bar{\epsilon}_t = \beta_t - \bar{\gamma}_t, \qquad f_t(x) = f\big(\sigma_t, \gamma_t(x), \gamma_t''(x)\big), \tag{6.15}$$

with f given by (6.2). Then inequality (1.33) is satisfied if and only if

$$0 \le l\sigma_t\dot{\beta}_t - \dot{E}_t = -\int_0^l f_t(x)\dot{\gamma}_t(x)dx.$$

By the dissipation inequality (5.4) and the equilibrium condition (6.3), this condition is verified if and only if

$$f_t(x)\dot{\gamma}_t(x) = 0 \quad \text{a.e. } x \in (0, l), \tag{6.16}$$

where *a.e.* means *except at a finite or countable number of points*. Indeed, f_t depends on the second derivative γ_t'', which may not exist at a finite number of points.

Like in the local model, a *quasi-static evolution* is a deformation process made of stable equilibrium configurations, and the quasi-static evolution corresponding to a given load process $t \mapsto \beta_t$ can be determined by incremental energy minimization. For this, with the due changes, we repeat the procedure followed in Sect. 5.4 for the local model. For the energy

$$E_t(\gamma) = l w(\beta_t - \bar{\gamma}) + \int_0^l \Big(\theta(\gamma(x)) + \frac{1}{2}\alpha\gamma'^2(x)\Big)dx,$$

we consider the expansion (5.24), with $\gamma = \gamma_t + \tau\dot{\gamma}$ and

$$I_t(\dot{\gamma}) = l\sigma_t\dot{\beta}_t + \int_0^l f_t(x)\dot{\gamma}(x)dx,$$
$$J_t(\dot{\gamma}) = lw''(\bar{\epsilon}_t)\big(\dot{\beta}_t^2 - 2\dot{\beta}_t\bar{\dot{\gamma}}\big) + \int_0^l \big(w''(\bar{\epsilon}_t)\bar{\dot{\gamma}}^2 + \theta''\big(\gamma_t(x)\big)\dot{\gamma}^2(x) + \alpha\dot{\gamma}'^2(x)\big)dx, \tag{6.17}$$

and we minimize the second-order approximation

$$F_t(\dot{\gamma}) = I_t(\dot{\gamma}) + \frac{1}{2}\tau J_t(\dot{\gamma}).$$

Note that, while in the local model the $\dot{\gamma}$ are constants, here they are functions satisfying the constraint $\dot{\gamma}(x) \geq 0$ and the boundary conditions

$$\dot{\gamma}(0) = 0, \qquad \dot{\gamma}(l) = 0. \tag{6.18}$$

A necessary condition for a minimum at $\dot{\gamma} = \dot{\gamma}_t$ is that the first variation

$$\delta F_t(\dot{\gamma}_t, \delta\dot{\gamma}) = \int_0^l f_t(x)\delta\dot{\gamma}(x)dx - \tau w''(\bar{\epsilon}_t)\dot{\beta}_t \int_0^l \delta\dot{\gamma}(x)dx$$
$$+ \tau \int_0^l \big(w''(\bar{\epsilon}_t)\bar{\dot{\gamma}}_t + \theta''\big(\gamma_t(x)\big)\dot{\gamma}_t(x) - \alpha\dot{\gamma}_t''(x)\big)\delta\dot{\gamma}(x)dx$$

be non-negative for all perturbations $\delta\dot{\gamma}$ which satisfy the boundary conditions

$$\delta\dot{\gamma}(0) = \delta\dot{\gamma}(l) = 0, \tag{6.19}$$

and the compatibility condition

$$\dot{\gamma}_t(x) + \delta\dot{\gamma}(x) \geq 0 \quad \forall x \in (0, l), \tag{6.20}$$

which follow from (6.7) and $(5.23)_2$, respectively. Because

$$\dot{f}_t(x) = \theta''\big(\gamma_t(x)\big)\dot{\gamma}_t(x) - w''(\bar{\epsilon}_t)(\dot{\beta}_t - \bar{\dot{\gamma}}_t) - \alpha\dot{\gamma}_t''(x), \tag{6.21}$$

the non-negativeness of the first variation is expressed by

$$\int_0^l \big(f_t(x) + \tau\dot{f}_t(x)\big)\delta\dot{\gamma}(x) \geq 0. \tag{6.22}$$

Therefore, a necessary condition for a minimum is that

$$\big(f_t(x) + \tau\dot{f}_t(x)\big)\delta\dot{\gamma}(x) \geq 0 \quad \text{a.e. } x \in (0, l). \tag{6.23}$$

By (6.20), at points at which $\dot{\gamma}_t(x) > 0$ the perturbation $\delta\dot{\gamma}(x)$ may take any sign, and therefore the term within parentheses must be zero. At points at which $\dot{\gamma}_t(x) = 0$ the perturbation is arbitrary positive, and therefore the same term must be non-negative. This determines the Kuhn-Tucker conditions

$$\dot{\gamma}_t(x) \geq 0, \quad f_t(x) + \tau \dot{f}_t(x) \geq 0, \qquad \big(f_t(x) + \tau \dot{f}_t(x)\big)\dot{\gamma}_t(x) = 0, \tag{6.24}$$

associated with the minimum problem.

In particular, in the inelastic zone $f_t(x)$ is zero by definition. Therefore, conditions (6.24) reduce to

$$\dot{\gamma}_t(x) \geq 0, \qquad \dot{f}_t(x) \geq 0, \qquad \dot{f}_t(x)\dot{\gamma}_t(x) = 0. \tag{6.25}$$

The complementarity condition then poses the alternative between $\dot{\gamma}_t(x) = 0$ and $\dot{f}_t(x) = 0$. By (6.21), the second option requires that $\dot{\gamma}_t$ be a solution of the differential equation

$$\theta''\big(\gamma_t(x)\big)\dot{\gamma}_t(x) - w''(\bar{\epsilon}_t)(\beta_t - \bar{\dot{\gamma}}_t) - \alpha \dot{\gamma}_t''(x) = 0. \tag{6.26}$$

We say that $\dot{\gamma}_t$ is a *localized continuation* if $\dot{\gamma}_t(x) = 0$ on some subinterval of the inelastic zone, and that it is a *full-size continuation* otherwise. We also call *localization zone* the set at which Eq. (6.26) holds. For a full-size continuation the localization zone coincides with the inelastic zone, while for a localized continuation it is a proper subset of the inelastic zone. In this second case, there is at least one interval at which

$$\dot{f}_t(x) = -w''(\bar{\epsilon}_t)(\beta_t - \bar{\dot{\gamma}}_t) = -\dot{\sigma}_t. \tag{6.27}$$

By $(6.25)_2$, this is impossible in a work-hardening response, $\dot{\sigma}_t > 0$. By consequence, *there are no localized continuations for a hardening response.*

Due to its quadratic structure, the functional F_t admits the finite expansion

$$F_t(\dot{\gamma}_t + \delta\dot{\gamma}) = F_t(\dot{\gamma}_t) + \delta F_t(\dot{\gamma}_t, \delta\dot{\gamma}) + \frac{1}{2}\delta^2 F_t(\dot{\gamma}_t, \delta\dot{\gamma}),$$

where the second variation

$$\delta^2 F_t(\dot{\gamma}_t, \delta\dot{\gamma}) = \int_0^l \big(\theta''\big(\gamma_t(x)\big)\delta\dot{\gamma}^2(x) + w''(\bar{\epsilon}_t)\delta\bar{\dot{\gamma}}^2 + \alpha\delta\dot{\gamma}'^2(x)\big)dx, \tag{6.28}$$

is, in fact, independent of $\dot{\gamma}_t$. A sufficient condition for a local minimum at $\dot{\gamma}_t$ is that the first variation be non-negative, and the second variation be non-negative for all $\delta\dot{\gamma}$ for which the first variation is zero.

For perturbations with $\delta\gamma(x) > 0$ out of the inelastic zone, inequality (6.22) is strict for sufficiently small τ, and therefore the first variation is positive. Therefore, it is sufficient to consider perturbations with support in the inelastic zone. After observing that the second variation of F_t coincides with the right-hand side of (6.9) with $\delta\gamma$ replaced by $\delta\dot{\gamma}$, we conclude that a sufficient condition for a minimum at $\dot{\gamma}_t$ is the non-negativeness of the smallest eigenvalue of the eigenvalue problem (6.10), with $(a, a + l_j)$ the largest interval contained in the inelastic zone.

Moreover, if $\dot{\gamma}_t$ is a full-size continuation, $\dot{\gamma}_t(x)$ is positive almost everywhere in the inelastic zone. Hence, by (6.20), $\delta\dot{\gamma}(x)$ is not restricted in sign. Therefore, the eigenvalue problem refers to perturbations subject only to the boundary conditions (6.19). If $\dot{\gamma}_t$ is a localized continuation, by (6.27) $\dot{f}_t(x)$ is positive out of the localization zone, except in the singular case of a perfectly plastic response. Then, by (6.22), in a softening response the first variation is positive for perturbations with $\delta\gamma(x) > 0$ out of the localization zone. Therefore, a sufficient condition for a minimum is the non-negativeness of the smallest eigenvalue of problem (6.10), with $\delta\dot{\gamma}$ not restricted in sign and with support inside the localization zone. Clearly, this condition is also necessary, since a negative eigenvalue implies that the second variation evaluated at any corresponding eigenfunction is negative.

While for the local model the explicit determination of a quasi-static evolution was reduced to the integration of a differential equation, for the non-local model the task is not so easy. Only in some special cases, like at the onset of the inelastic regime discussed in the next Section, a solution to the incremental minimization problem can be determined in a closed form.

6.5 The Onset of the Inelastic Regime

Consider a load process $t \mapsto \beta_t$ from the natural configuration $\epsilon = \gamma = 0$, with $\dot{\beta}_t > 0$ for all t. Like in the local model, the evolution initially follows the elastic regime, in which γ_t is zero, ϵ_t is constant, and

$$\epsilon_t = \beta_t, \qquad \sigma_t = w'(\beta_t), \qquad f_t = \theta'(0) - \sigma_t > 0. \tag{6.29}$$

The elastic regime ends when β_t reaches the value $\beta_c = (w')^{-1}(\theta'(0))$ and, consequently, f_t becomes zero. The inelastic zone, which in the elastic regime was empty, suddenly becomes equal to $(0, l)$, and the inelastic regime begins. In this Section we determine the continuations $\dot{\gamma}_t$ at the onset of the inelastic regime, that is, at $\beta = \beta_c$.

At the onset, the differential equation (6.26) has the special form

$$\theta''(0)\dot{\gamma}(x) - w''(\beta_c)(\dot{\beta} - \bar{\dot{\gamma}}) - \alpha\dot{\gamma}''(x) = 0, \tag{6.30}$$

where, for simplicity, the subscript t has been omitted. We initially assume that $\dot{\gamma}$ is a full-size continuation. Then Eq. (6.30) holds at all x in $(0, l)$, and we solve it for the boundary conditions (6.18). The solution depends on the sign of $\theta''(0)$. For $\theta''(0)$ positive, the solution is

$$\dot{\gamma}(x) = \frac{w''(\beta_c)(\dot{\beta} - \bar{\dot{\gamma}})}{\theta''(0)}\left(1 - \frac{\cosh\kappa(l/2 - x)}{\cosh\kappa l/2}\right), \tag{6.31}$$

with $\kappa = (\theta''(0)/\alpha)^{1/2}$. By integration over $(0, l)$, one has

$$\bar{\dot{\gamma}} = \frac{w''(\beta_c)(\dot{\beta} - \bar{\dot{\gamma}})}{\theta''(0)}\varphi(\kappa l), \qquad \varphi(\kappa l) = 1 - \frac{\tanh\kappa l/2}{\kappa l/2}, \tag{6.32}$$

that is,

$$\bar{\dot{\gamma}} = \frac{\varphi(\kappa l)w''(\beta_c)}{\theta''(0) + \varphi(\kappa l)w''(\beta_c)}\dot{\beta}. \tag{6.33}$$

A comparison with the inelastic response in Eq. (5.28) shows that the non-local effect is given by the factor $\varphi(\kappa l)$. Because $0 < \varphi(\kappa l) < 1$, from the preceding equation we deduce that $\bar{\dot{\gamma}} < \dot{\beta}$, and from (6.31) it follows that the dissipation inequality $\dot{\gamma}(x) \geq 0$ is satisfied everywhere. Finally, using the relation $\dot{\sigma} = w''(\beta_c)(\dot{\beta} - \bar{\dot{\gamma}})$, the incremental force-elongation relation

$$\dot{\sigma} = \frac{\theta''(0)w''(\beta_c)}{\theta''(0) + \varphi(\kappa l)w''(\beta_c)}\dot{\beta} \tag{6.34}$$

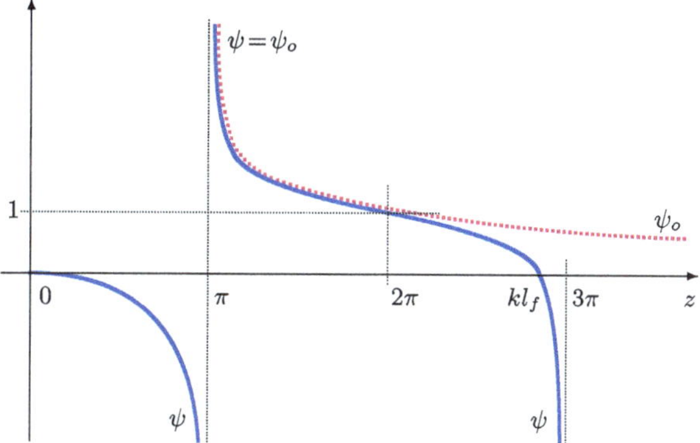

Fig. 18 The functions ψ (*solid line*) and ψ_o (*dotted line*), as defined in $(6.36)_2$ and (6.13), respectively

is obtained. It shows that the slope $\dot{\sigma}/\dot{\beta}$ of the response curve (σ, β) is positive, larger than the slope (5.29) in the local model, but smaller than the slope $w''(\beta_c)$ of the elastic response.

The singular case $\theta''(0) = 0$ will not be considered for brevity. The interested reader is addressed to [56]. For $\theta''(0) < 0$, the solution of the differential problem (6.30), (6.18) is

$$\dot{\gamma}(x) = \frac{w''(\beta_c)(\dot{\beta} - \dot{\bar{\gamma}})}{\theta''(0)} \left(1 - \frac{\cos k(l/2 - x)}{\cos kl/2}\right), \tag{6.35}$$

with $k = (-\theta''(0)/\alpha)^{1/2}$. Integration over $(0, l)$ yields

$$\dot{\bar{\gamma}} = \frac{w''(\beta_c)(\dot{\beta} - \dot{\bar{\gamma}})}{\theta''(0)} \psi(kl), \qquad \psi(z) = 1 - \frac{\tan z/2}{z/2}, \tag{6.36}$$

so that

$$\dot{\bar{\gamma}} = \frac{\psi(kl) w''(\beta_c)}{\theta''(0) + \psi(kl) w''(\beta_c)} \dot{\beta}, \tag{6.37}$$

and

$$\dot{\sigma} = \frac{\theta''(0) w''(\beta_c)}{\theta''(0) + \psi(kl) w''(\beta_c)} \dot{\beta}. \tag{6.38}$$

Again, the difference with (5.28) and (5.29) is due to a single factor, $\psi(kl)$. The graph of ψ is shown in Fig. 18. For $\pi < kl < 2\pi$, ψ coincides with the function ψ_o defined in (6.13). For $kl < \pi$, $\psi(kl)$ is negative. Then, because $\theta''(0) < 0$, the denominator in (6.37) is negative. Then $\dot{\bar{\gamma}}$ is positive and less than $\dot{\beta}$, and from (6.35) it follows that $\dot{\gamma}(x)$ is positive for all x. Therefore, the function (6.35) is indeed a continuation if $kl < \pi$.

At $kl = \pi$, ψ jumps from $-\infty$ to $+\infty$. For $kl > \pi$, ψ is positive and decreasing, and becomes zero at $kl = 2.861\pi$. The denominator of (6.37) is positive for kl close to π and decreases with increasing kl. It becomes zero at the critical value $kl_f \in (\pi, 2.861\pi)$ at which

$$\psi(kl_f) = \psi_f, \qquad \psi_f = -\frac{\theta''(0)}{w''(\beta_c)}. \tag{6.39}$$

Then $\dot{\bar{\gamma}}$ is positive for $\psi(kl) > \psi(kl_f)$, that is, for $kl < kl_f$. Moreover, by (6.35), $\dot{\gamma}(x)$ is positive at all x only if $kl \leq 2\pi$. Indeed, for $kl > 2\pi$, $\dot{\gamma}(x)$ is negative near the boundary.

Therefore, for $kl > \pi$ the full-size solution (6.35) is admissible only for

$$kl < \min\{kl_f, 2\pi\}. \tag{6.40}$$

In the remaining case $2\pi < kl < kl_f$, no full-size solution is possible. Then we consider localized continuations, that is, functions $\dot{\gamma}(x)$ which are positive on an interval $(a, a+l_i)$, with $a \geq 0$, $a+l_i \leq l$, and $l_i < l$, and zero outside. For such continuations, the differential equation (6.30) holds only in the interval $(a, a+l_i)$, and the boundary conditions are

$$\dot{\gamma}(a) = 0, \qquad \dot{\gamma}(a+l_i) = 0. \tag{6.41}$$

Indeed, these are the end conditions (6.18) if $a = 0$ and if $a + l_i = l$, respectively. Otherwise, the same conditions are imposed by the continuity with the neighboring region at which $\dot{\gamma}(x) = 0$. Moreover, since $l_i < l$, at least one of the points $a, a+l_i$ is an interior point of $(0, l)$. Because $\dot{\gamma}'$ is supposed to be continuous, by continuity with the neighboring region at which $\dot{\gamma}'(x) = 0$, at least one of the conditions

$$\dot{\gamma}'(a) = 0, \qquad \dot{\gamma}'(a+l_i) = 0, \tag{6.42}$$

must be added to the problem. This brings the supplementary condition

$$kl_i = 2\pi, \tag{6.43}$$

which determines the length l_i of the localization zone. Thus, l_i does not depend on the length of the bar, but only on the material constant k. The solution of the differential problem is

$$\dot{\gamma}(x) = \begin{cases} \frac{\bar{\dot{\gamma}} - \dot{\beta}}{\psi_f}(1 - \cos k(x-a)) & \text{if } a < x < a + 2\pi/k, \\ 0 & \text{otherwise,} \end{cases} \tag{6.44}$$

with ψ_f as in $(6.39)_2$. Integrating over $(a, a+2\pi/k)$ we find

$$\bar{\dot{\gamma}} = \frac{2\pi}{kl}\frac{\bar{\dot{\gamma}} - \dot{\beta}}{\psi_f}, \tag{6.45}$$

that is,

$$\bar{\dot{\gamma}} = \frac{1}{1 - \frac{kl}{2\pi}\psi_f}\dot{\beta}. \tag{6.46}$$

The denominator is positive if

$$kl < \frac{2\pi}{\psi_f} = kl_f. \tag{6.47}$$

In this case $\bar{\dot{\gamma}}$ is positive and, by (6.44), $\dot{\gamma}(x)$ is positive at all x.

Using the definitions (6.13), $(6.36)_2$, $(6.39)_2$ of ψ_o, ψ and ψ_f, the incremental force-elongation relation can be given the form

$$\dot{\sigma} = \begin{cases} \frac{\psi_f w''(\beta_c)}{\psi_f - \psi(kl)}\dot{\beta} & \text{if } kl < 2\pi, \\ \frac{\psi_f w''(\beta_c)}{\psi_f - \psi_o(kl)}\dot{\beta} & \text{if } kl > 2\pi. \end{cases} \tag{6.48}$$

The analysis of the slope $\dot{\sigma}/\dot{\beta}$ of the response curve (σ, β) shows that the response at the onset of the inelastic deformation is work-hardening if $kl < \pi$ and strain-softening if $kl > \pi$. Therefore, for $\theta''(0) < 0$, at the onset of the inelastic regime there are three types of continuations:

(i) the full-size hardening continuations (6.35) for $kl < \pi$,
(ii) the full-size softening continuations (6.35) for $\pi < kl < \min\{kl_f, 2\pi\}$,
(iii) the localized softening continuations (6.44) for $kl_f > 2\pi$ and $2\pi < kl < kl_f$.

Equation (6.48) also shows that for $\psi_o(kl)$ approaching ψ_f from above the slope $\dot{\sigma}/\dot{\beta}$ tends to $-\infty$. That is, the equality $\psi_o(kl) = \psi_f$ corresponds to catastrophic fracture. In particular, the occurrence of this event at the onset of the inelastic regime corresponds to the totally brittle fracture of Griffith's theory, reproduced by the local model. Note that the denominator of (6.48) can be zero only if $kl > \pi$, since $\psi(kl) < 0$ for $kl < \pi$. Therefore, *there is no catastrophic fracture at the onset for* $kl < \pi$.

For $\psi_o(kl) > \psi_f$, at the onset an inelastic regime begins. The determination of the continuations and the characterization of catastrophic fracture are more complicated inside the inelastic regime. Indeed, just after the onset, $\theta''(\gamma(\cdot))$ ceases to be constant, and the possibility of a closed form determination of $\dot{\gamma}_t$ vanishes. By consequence, the only effective way for characterizing catastrophic failure at inelastic response is by numerical simulation. In fact, this has been done in the first numerical simulation reported in Sect. 6.7 below.

6.6 Internal Lengths

It is commonly said that the description of softening response is made possible by the introduction of an *internal length* of the material, and that this is precisely the role played by the additional gradient term of the non-local model. Here we show that, in fact, *two internal lengths* are required for an adequate description of the response in the non-local model. Moreover, these internal lengths are not really material constants, since their values may vary during a quasi-static evolution.

Consider first the onset of the inelastic deformation. As shown in the previous Section, the response at the onset is determined by the length l of the bar and by three material constants:

- *the initial curvature* $\theta''(0)$ *of the cohesive energy,*
- *the Young modulus* $w''(\beta_c)$ *at the onset,*
- *the non-locality parameter* α.

They naturally combine to form the ratios

$$l_i = \frac{2\pi}{k} = 2\pi\sqrt{\frac{\alpha}{-\theta''(0)}}, \qquad l_c = 2\pi\sqrt{\frac{\alpha}{w''(\beta_c)}}, \tag{6.49}$$

both with the physical dimension of a length. They will be called the *internal length* and the *characteristic length* of the material at the onset, respectively. The length l_f defined in (6.39) is the function of l_i and l_c defined by $\psi(kl_f) = \psi_f = (l_c/l_i)^2$.

An illustration of the way in which l_i and l_c determine the response at the onset is given in Fig. 19. In the plane $(l/l_c, l/l_i)$, each point represents a bar. The curve Ψ is the set of all bars which are in the limit situation $\psi_o(kl) = \psi_f$. With (6.49), this equation takes the form

$$\psi_o\left(\frac{2\pi l}{l_i}\right) = \left(\frac{l_c}{l_i}\right)^2, \tag{6.50}$$

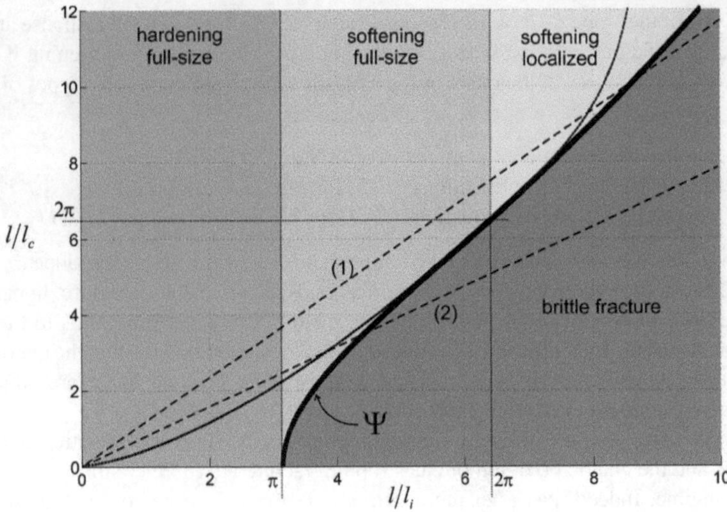

Fig. 19 Diagram of the response at the onset of the inelastic regime, as determined by the internal length l_i and by the characteristic length l_c

and the equation of the curve is

$$\frac{l}{l_c} = \frac{l/l_i}{\sqrt{\psi_o(2\pi l/l_i)}} \doteq \Psi(l/l_i). \tag{6.51}$$

The points located below the curve represent the bars which undergo totally brittle fracture at the onset. For every value of l/l_i, the curve provides the minimum value of l/l_c for which a bar is safe from totally brittle fracture at the onset.

The three vertical strips correspond to the solutions of type (i)–(iii) in the preceding Section. They show how the ratio l/l_i controls the quality of the incremental response, hardening or softening, full-size or localized. The straight lines (1), (2) represent two families of bars, each with the same ratio l_c/l_i, that is, made of the same material, with increasing length starting from $l = 0$ at the origin. The picture shows that for small l the response is work-hardening and full-size, and for large l there is catastrophic rupture at the onset. For intermediate values, the response is strain-softening. More precisely, if $l_i < l_c$ as in the family (2) the softening response is full-size, while for $l_i < l_c$ as in the family (1) the response is full-size for $l < 2\pi l_i$ and localized for $l > 2\pi l_i$.

No such precise conclusions can be drawn for the subsequent evolution. However, the results obtained for the onset throw some light on some general qualitative aspects of the inelastic response. Indeed, at least in principle, internal length and characteristic length can be defined not only at the onset, but at any stage of the inelastic regime, and it is reasonable to believe that they determine the continuations in the same way as they do at the onset.

Let a *current internal length* and a *current characteristic length* be defined as in (6.49), with $\theta''(0)$ replaced by an appropriate average of $\theta''(\gamma_t(x))$ and $w''(\beta_c)$ replaced by $w''(\bar{\epsilon}_t)$. In the plane of Fig. 19 the bar is now represented by a point which moves with t, and the fracture curve also varies with t. Let us assume that θ'' varies more rapidly than w'', so that the point describes an almost horizontal line.

Assume that θ is concave and that θ' is convex. Then θ'' is negative and monotonic increasing and, since $t \mapsto \gamma_t(x)$ is increasing due to the dissipation inequality, $t \mapsto \theta''(\gamma_t(x))$

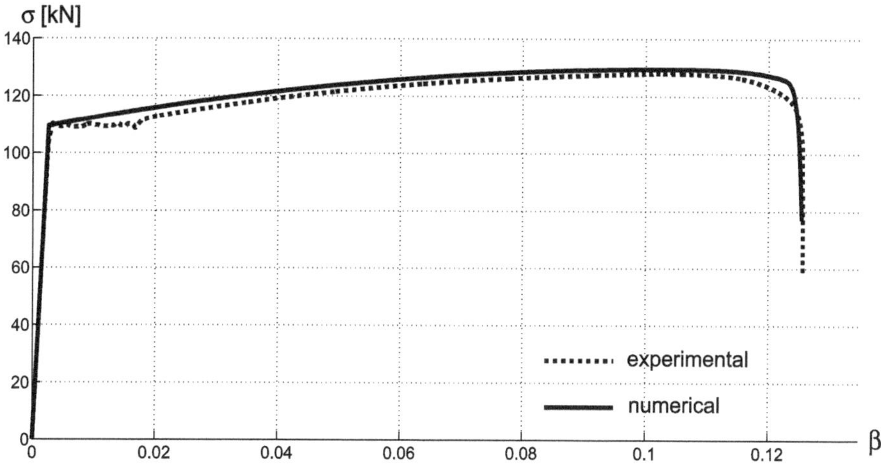

Fig. 20 Force-elongation response curves for a steel bar: experimental (*black solid line*), and numerical (*dotted line*). The numerical curve has been obtained with a convex-concave cohesive energy. The curve obtained with a concave-convex-concave cohesive energy almost coincides with the experimental curve

is increasing at all x. Then the average of $\theta''(\gamma_t(x))$ increases with t, and therefore the internal length increases as well. In the picture, the point representing the beam point moves to the right, and when it reaches the curve Ψ, catastrophic fracture occurs.

On the contrary, if θ' is concave the point moves to the left, that is, towards the region in which the slope of the response curve tends to zero. Thus, the following qualitative conclusion emerges: *a cohesive energy with convex θ' favors brittle fracture, and a cohesive energy with concave θ' favors ductile fracture.*

6.7 Numerical Simulations

To give an idea of the effectiveness and flexibility of the non-local model, let us briefly illustrate the results of two numerical simulations taken from [56]. Their purpose was to reproduce the experimental (σ, β) response curves of a steel bar and of a concrete specimen, respectively.

In both experiments, the load consists of a monotonically increasing displacement applied to one of the end sections. In the simulations, the elastic energy density is assumed to be quadratic

$$w(\epsilon) = \frac{1}{2} E A \epsilon^2,$$

and the cohesive energy density θ is assumed to be strictly increasing, piecewise cubic, and C^2.

The experimental curve of the steel bar, dotted line in Fig. 20, exhibits an initial linear elastic behavior, followed by a horizontal plateau, a work-hardening regime, and a softening regime ending up in catastrophic rupture. This response is well described by a convex-concave cohesive energy, in which the initial convexity reproduces the hardening part of the response diagram, and the concavity reproduces the softening regime and the final rupture.

The computed response curve is the solid line of Fig. 20. The agreement with the experimental curve is very good, except at the horizontal plateau. To improve the model's response

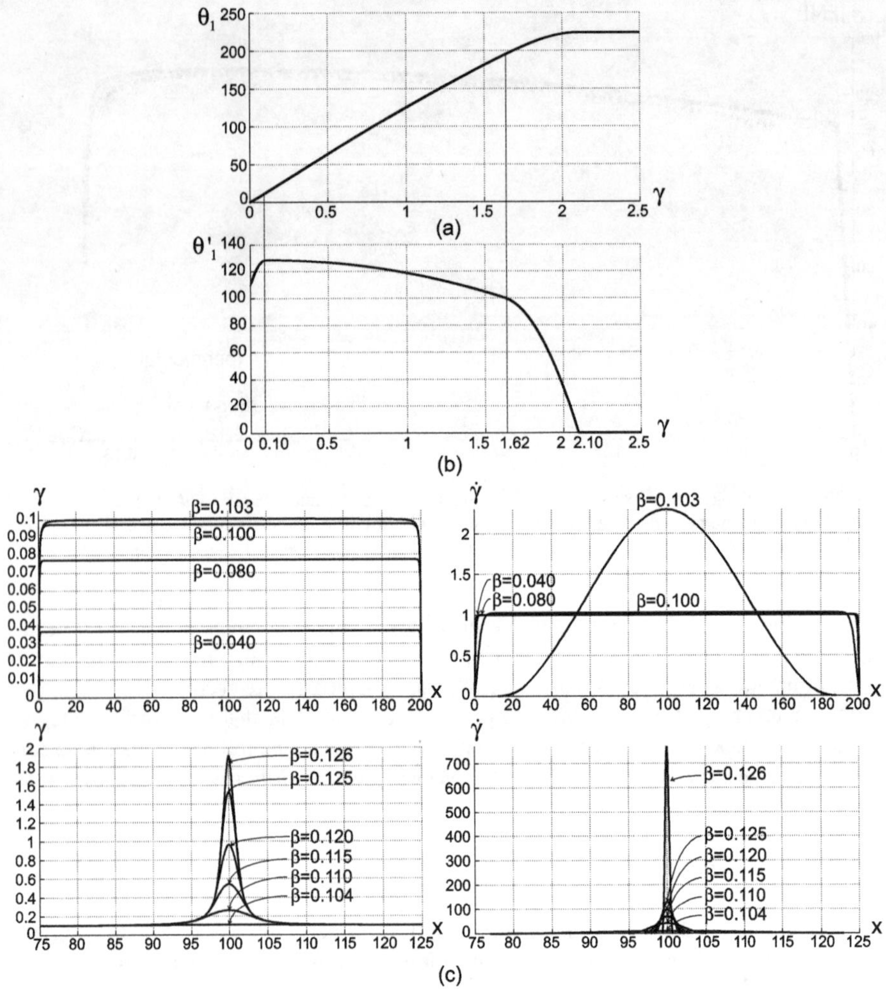

Fig. 21 The cohesive energy assumed for the steel bar (**a**), its derivative (**b**), and the evolution of the inelastic deformation γ and of its derivative $\dot{\gamma}$ (**c**)

in this specific zone, it was sufficient to slightly modify the expression of θ by adding a small initial concavity. The modified response curve can hardly be distinguished from the experimental one. As said above, a C^2 and piecewise cubic representation was taken for θ. Its original expression involved four material constants, to be determined by fitting the experimental curve. For the improved simulation reproducing the plateau, six constants have been employed.

The assumed shapes of θ and of its derivative θ' in the four-constants simulation are shown in Figs. 21(a) and 21(b). The initial convexity of θ is too mild to be detected in (a), and is revealed only by the short initial ascending branch of θ' in (b). Even less evident are the effects, not shown in the picture, of the initial concavity added in the six-constants simulation. This shows that some macroscopically evident aspects of the response, such as

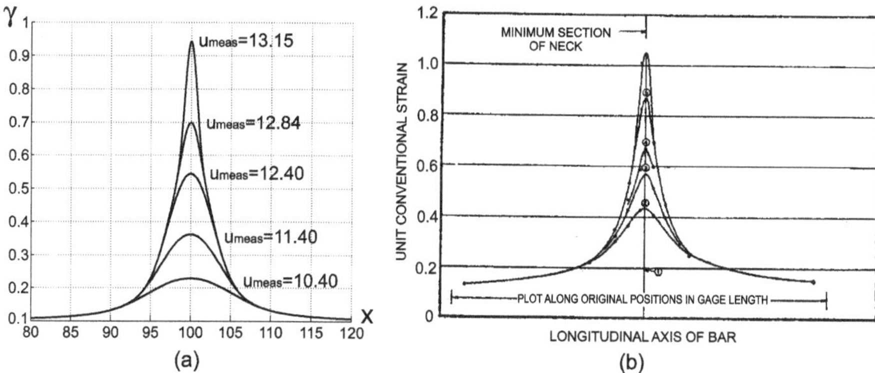

Fig. 22 The evolution of the inelastic deformation in the softening regime (**a**), compared with the experimental curves of [106] (**b**)

the initial plateau and the initial hardening regime, depend on fine properties of the cohesive energy which can hardly be perceived in a graphical representation of θ.

The evolution of the inelastic deformation is shown in Fig. 21(c). Localization starts just after the force σ attains a maximum, and grows progressively up to ultimate rupture. Of interest is the comparison, made in Fig. 22, with the experimental curves of [106], which date back to the 1940's, long before the appearance of theories of softening and strain localization.

For the concrete specimen, the experimental response curve is the dotted line in Fig. 23(a). After reaching a maximum the force decreases, and eventually tends to a positive asymptotic limit. In the numerical simulation, this curve is reproduced taking a convex-concave energy which, at a first sight, does not differ much from the energy shown in Fig. 21(a). The difference is that the derivative θ', instead of being concave as in Fig. 21(b), is convex, except in the vicinity of the origin. This determines the convex shape of the descending part of the response curve, and leads to a ductile fracture mode, in accordance with the theoretical prediction made in Sect. 6.6.

Just after the response curve reached a maximum, the inelastic strain localizes. As shown by the strain profiles in Fig. 23(b), the inelastic zone initially shrinks, and then, in correspondence with the decreasing part of the response curve, it expands again. Simulations not reported in the figure show that for further increasing load the inelastic region progressively expands, and eventually spreads over the whole bar.

6.8 References and Comments

A great progress in the understanding of material response came from the study of the phenomenon of softening. Initially this phenomenon was considered anomalous, and Drucker's postulate [64] seemed to exclude it, once for all, from the domain of the mathematically well-posed and physically plausible theories. But its systematic appearance in the response of non-metallic materials soon suggested to re-consider this position.

Also, due to the strong dependence of the results on the mesh size, softening appeared to be intractable with ordinary finite element techniques. A remedy devised by many researchers was to smear the cracks over finite volume portions. By the mid 1980's, a huge amount of literature on models with distributed cracking was available. A detailed picture of the state of the art can be found in the review paper [12].

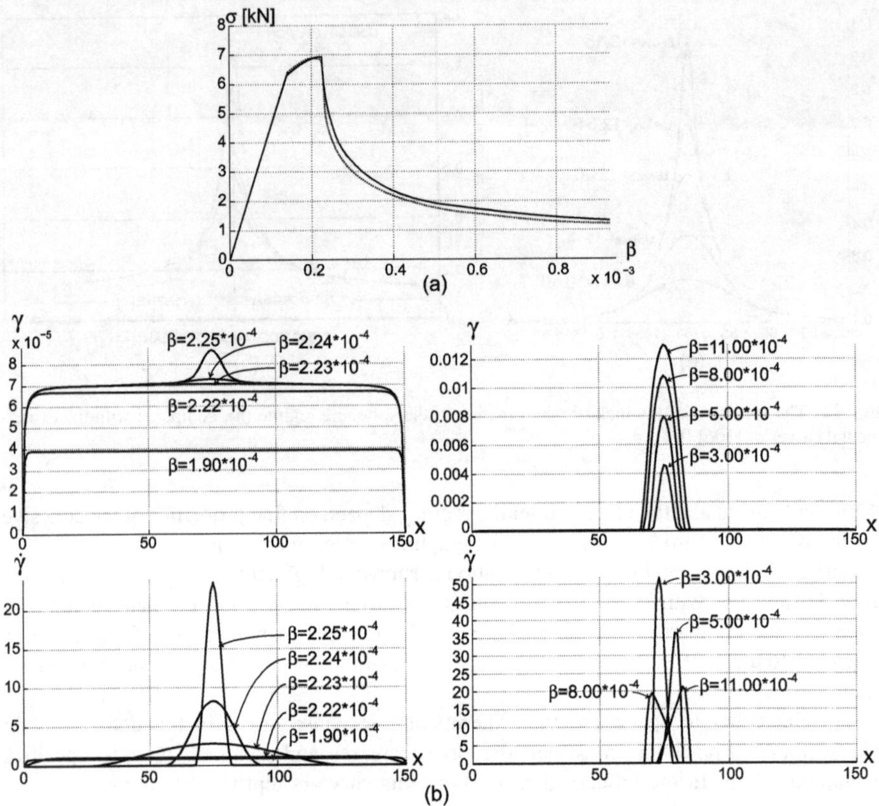

Fig. 23 Force-elongation response curves for a concrete specimen: experimental (*solid line*), and numerical (*dotted line*) (**a**), and the evolution of the inelastic deformation γ and of its derivative $\dot{\gamma}$ (**b**)

To overcome the difficulties in the numerical treatment, a decisive step was the introduction of non-local terms in the governing equations. This was done in two different ways, either by considering generalized continua with microstructure or, in the more traditional domain of ordinary continua, by adding terms involving gradients of the state variables. The two options are analyzed in the papers [90] and [91], respectively, which also provide a history of the subject. Additional references can be found in the paper [48].

To the writer's knowledge, the idea of using gradient energy terms to model softening phenomena traces back to Aifantis [2]. In fact, this is not the first time in which gradient terms drastically improved the range of a mechanical model. This had occurred before, with Maxwell's theory for rarefied gases [103], with Van der Waals' [37, 130, 131] and Korteweg's [92] theories for capillarity, Cahn and Hilliard's model for phase transition in metals [29], Mumford and Shah's image segmentation theory [107], with theories of liquid crystals [67, 70], and with numerical simulations for the rupture of concrete [49]. In the variational approach to fracture, the introduction of gradient terms is relatively recent. A precursor was the paper [6] of Ambrosio and Tortorelli, whose original goal was to provide a mathematically sound algorithm for the problem of image segmentation. Its application to fracture by Bourdin [23] was followed by a number of explicit numerical solutions to fracture problems [7, 22, 55, 93].

Mechanical models for damage involving gradient terms have been proposed and discussed by several authors [17, 71, 91, 98, 115]. The model proposed in the paper [56] and presented in this Lecture is a special case, in which the dependence of the energy density on the elastic and inelastic parts of the deformation is uncoupled. This rules out the decay of the inelastic modulus predicted by damage theories, and the result is a response typical of plasticity. However, the analysis of the response at the onset reported in Sect. 6.5 can be easily extended to damage models. Indeed, since at the onset the inelastic deformation is zero, the first step of the incremental energy minimization is the same for damage and plasticity. With appropriate changes, the qualitative analysis of the subsequent evolution can be made along the lines followed in Sect. 6.6 for plastic response.

A natural direction for future work is to extend the model to higher dimension. Due to substantial mathematical difficulties, this task is far from trivial. Another challenging perspective is to find a way to correlate the analytical properties of the cohesive energy, namely, concavity or convexity of θ and θ', to the microscopic properties of real materials. This problem goes far beyond classical continuum mechanics, and requires an interdisciplinary approach involving several research fields in physics, chemistry, and materials science.

We close with some technical comments on the non-local model presented in this Lecture.

1. It is well known that the introduction of a gradient term acts as a *localization limiter* [15, 96]. That is, it prevents an immediate extreme localization in a softening response. For the present model, this stabilizing effect has been pointed out at the end of Sect. 6.3. In the analysis of the response at the onset, Sect. 6.5, it has been found that the length of the localized zone at the onset is a material constant, the *internal length* l_i. In the subsequent extrapolation to the incremental response in the inelastic regime, Sect. 6.6, it has been conjectured that the internal length is a parameter that evolves in a way depending on the shape of the cohesive energy density θ and on the current inelastic deformation γ_t.

2. In the present one-dimensional model, the *process zone* observed in ductile fracture is represented by the *inelastic zone*. This zone appears at the onset of the inelastic deformation. In the subsequent evolution, its size can grow or decrease, depending on the convexity-concavity properties of θ'. In spite of some early realization [81] and of some more recent observations on the dependence of the response curve on the shape of the damage law, see for example Fig. 16 of [91], the correlation between fracture modes and the convexity-concavity properties of θ' did not receive adequate attention in the literature.

3. In the numerical simulations, a quasi-static evolution is determined by the minimization of the second-order approximation of the functional $E_{t+\tau}$, restricted to a set of stable equilibrium configurations. In this set, the functional is positive. For this reason, no problem of mesh dependence has been met in the numerical treatment.

4. Our choice of the boundary conditions (6.6) has been motivated in Sect. 6.3. Other authors [17, 91, 98, 115] prefer the natural conditions (6.5), mostly taken as equalities. A careful analysis of the two possibilities is made in [91]. Generally speaking, a point in favor of the natural boundary conditions is that they allow for homogeneous solutions, which have the advantage of being available in a closed form in the whole inelastic regime and not only at the onset [115]. On the other side, they can hardly be compared with experiments, since, as discussed in Sect. 6.2, it is very difficult to reproduce the natural boundary conditions in a real test.

5. At the onset of the inelastic regime, the constants θ''_J, $\theta''_{\min}$ which appear in the stability conditions (6.11), (6.14) are both equal to $\theta''(0)$, and the length l_J is equal to l. Then the necessary condition (6.11) reduces to $\alpha\lambda_l^2 + \theta''(0) \geq 0$, that is, to $\lambda_l^2 \geq k^2$, and the sufficient condition (6.14) reduces to the strict inequality $\lambda_l^2 > k^2$. In [56] it has been shown that

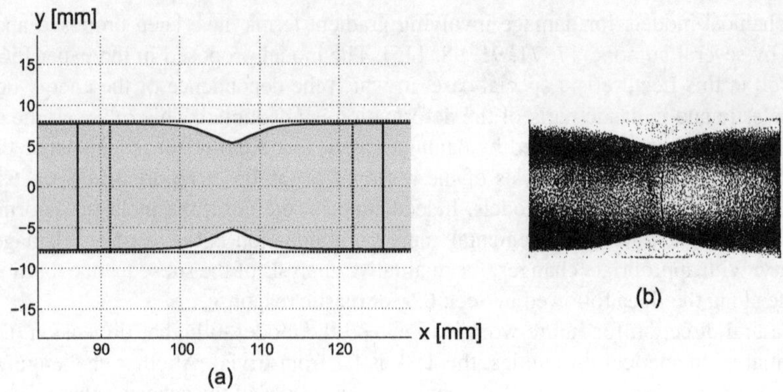

Fig. 24 The deformation profile of a bar just before fracture (**a**) compared with a picture taken from [106] (**b**)

these conditions coincide with $\psi_o(kl) \geq \psi_f$ and to the corresponding strict inequality, respectively. Therefore, catastrophic fracture and loss of stability come at the same time. This confirms the idea, already commented in Sect. 2.7, that fracture is in fact a special form of instability of equilibrium.

6. For the bar in the first simulation of Sect. 6.7, the deformation profile just before fracture is shown in Fig. 24(a). This profile has been deduced from the computed longitudinal deformation, multiplied by a coefficient of transverse contraction (Poisson ratio) equal to 0.3. There is a very good qualitative agreement with the experimental profile of Fig. 24(b), taken from [106], which shows the *necking* observed in the proximity of the rupture of metallic materials. It is remarkable that the present one-dimensional model provides an indirect measure of necking, which is intrinsically a multi-dimensional phenomenon.

7. The evolution curves of the inelastic deformation found in the simulation on the steel bar, Fig. 21(c), are very similar to those of [3, Fig. 6], which show the distribution of the shear strain in shear bands, and to the curves in [73] and in [91, Fig. 17], obtained more recently using both linear and exponential damage laws.

8. In [90, 91], the expansion of the inelastic zone observed in our second simulation is considered spurious, because it is accompanied by *stress locking*. This is the property that, in the force-elongation response curve (σ, β), σ tends to a positive limit when $\beta \to +\infty$, instead of decaying to zero as required by physical plausibility and confirmed by experiments [39, 132]. In fact, as shown in Fig. 3 of [89], in some models the limit force can be an important fraction, of the order of 80 %, of the maximum σ. In our second simulation, the positive limit for σ shown in Fig. 23(a) is not too large. It is probably due to the assumed constancy of the material parameter α. It does not seem impossible to obtain an expansion of the inelastic zone without stress locking, as required by the experiments. Anyway, this is a delicate point which deserves further study.

9. Another delicate point is the expansion of the inelastic zone found in the numerical simulations for the ductile fracture mode. This expansion agrees with the evolution curves obtained in several models [48, 110, 116], but is considered spurious in [90, 91]. There are good reasons for this. Indeed, this expansion seems to contradict the accessibility condition (6.16) which says that, as a consequence of the energy inequality (1.33), the inelastic deformation cannot increase at points at which $f_t(x) > 0$.

In fact, the contradiction is only apparent. Indeed, let the inelastic zone be $(a, a + l_j)$. Assume that, at $x = a$, γ_t'' jumps from zero to a positive value, as it occurs at the onset for the localized continuation (6.44). By (6.2), to a discontinuity of γ_t'' there corresponds a negative discontinuity

$$[\![f_t]\!](a) = -\alpha[\![\gamma_t'']\!](a)$$

of f_t. That is, f_t jumps from a positive value to zero, which is the value allowed in the inelastic region. As a consequence, in the accessibility condition (6.16) $f_t(a)$ must be replaced by the right limit $f_t(a+)$, which is zero. Therefore, there is no restriction on $\dot{\gamma}_t(a)$ coming from this condition.

Moreover, assume that the inelastic zone is expanding, $dx/dt < 0$ at $x = a$. Then by Hadamard's kinematic compatibility condition

$$[\![\dot{\gamma}_t']\!](x) = -[\![\gamma_t'']\!](x)\frac{dx}{dt},$$

(see, e.g., [77, Sect. 32.2]), a positive jump of γ_t'' is accompanied by a positive jump of $\dot{\gamma}_t'$. Then $\dot{\gamma}_t''(x)$ has a positive Dirac singularity and, by (6.21), $\dot{f}_t(x)$ has a negative singularity at $x = a$.

In the minimization of the second-order approximation of the energy, condition (6.16) is replaced by the milder complementarity condition

$$\big(f_t(x) + \tau\dot{f}_t(x)\big)\dot{\gamma}_t(x) = 0,$$

which reduces to (6.16) only if $\tau = 0$. But, due to the presence of the singularity of $\dot{f}_t$, condition (6.16) cannot be recovered in the limit for $\tau \to 0$. In the numerical simulation, the singularity of $\dot{f}_t$ smears over a rectangle $(x, x+\delta) \times (t, t+\tau)$ of the (x, t) plane, and in this rectangle condition (6.16) is replaced by the complementarity condition. This is the reason why the numerical procedure is able to capture the expansion of the inelastic zone in the ductile fracture mode.

7 Closure

In the last decades, a great progress in the understanding of fracture phenomena was achieved as a result of intense research activity, developed along several fronts: mathematics, engineering, numerics, experiment, and mechanical modeling. But, I believe, the progress was disordered. There was a big amount of duplication and dispersion, due to the lack of an adequate, universally accepted theoretical basis. The present Notes aim to be a small contribution to the clarification of some basic questions. Perhaps the reader has realized that, in spite of the apparent simplicity of the one-dimensional setting, the task is far from trivial.

I also believe in the opportunity of putting an end to the diaspora which subdivides the mechanics of materials into separate sections, each one with its own axioms, formalism, solution techniques, and jargon. The time has come for a unifying view, as it was our fathers' view when mechanics was in its flourishing youth. A modest contribution of these Notes is to show that plasticity, damage, and creation of microstructure have common theoretical bases, and can be treated in a unified way. And for a unified view I suggest incremental energy minimization as a basic analytical tool, and the representation of inelastic deformation by dissipative cohesive energy as a basic mechanical ingredient.

I have been frequently asked, whether other aspects of mechanical response can be included in the framework presented here. The only obstacle to an unconditioned positive answer is my poor knowledge of the many ongoing developments in the *mare magnum* of today's scientific literature. So, I just give a couple of examples of extensions which I consider as possible. Incremental energy minimization can be used successfully in problems of fatigue [88], and in special limit cases of material behavior such as no-tension materials [71]. Also, there are theories of growth and dissipation in biological tissues [4], which fit the scheme developed in Lectures 5 and 6 for plasticity.

I have also been asked whether dynamics can play any role in the framework described in the present Notes. Indeed, the framework adopted here is rigorously quasi-static, and looks self-consistent. Actually, there is at least a circumstance in which the present approach is inadequate. This is when the incremental minimization problem has no solution and, consequently, the system is forced to undergo a non-equilibrium transition. For example, this occurs at crack initiation in Lecture 1, and at the transition from a pre-fractured to a totally fractured regime in cases M and L of Lecture 2. This is a case in which a dynamic analysis could be helpful.

In a quasi-static analysis it is assumed that a non-equilibrium transition is instantaneous, and ends at a known equilibrium configuration, from which a quasi-static evolution re-starts. Of course, a different answer may come from a genuinely dynamic analysis. In this respect, a reassuring result comes from the study [66] on the one-dimensional problem of debonding of a thin film. The result is that, at least in the specific example considered there, the dynamic solution converges to the quasi-static solution when the loading velocity tends to zero. Unfortunately in the quasi-static context there are cases, see comment 4 in Sect. 2.7 above, in which the terminal equilibrium configuration of a non-equilibrium transition is not unique. In such cases, I see no alternative to a dynamic approach. For this, some recent attempts [25, 95] to extend to dynamics the Ambrosio-Tortorelli regularization technique could be very helpful.

I conclude with a short reference to some recent, very interesting research directions. In classical continuum mechanics, discontinuous displacement fields are considered as anomalous, and cause serious technical difficulties. This led to theories in which the differential equations are replaced by difference equations or by integral equations, as it occurs in *quantized fracture mechanics* [117, 118] and in *peridynamics* [122, 123], respectively. The new approaches naturally induce to a deeper insight into the physical structure of matter, and to a thorough revisitation of the classical model of the continuum. In this direction, the extensions of Griffith's and Barenblatt's model to fractal geometry [18, 35, 135] and the attempts of reconciling the continuum theories with atomistic simulations [74, 84] can be mentioned.

Trying to reduce to unity all these centrifugal research directions looks to be an almost desperate enterprise. The conclusion is that, in spite of the huge amount of existing literature, the study of fracture and other inelastic phenomena is still in its infancy. Reaching a reasonable, comprehensive theoretical scheme requires a long, collective, interdisciplinary effort.

References

1. Abeyaratne, R., Chu, C., James, R.D.: Kinetics of materials with wiggly energies: theory and application to the evolution of twinning microstructures in a Cu-Al-Ni shape memory alloy. Philos. Mag. A **73**, 457–497 (1996)
2. Aifantis, E.C.: Maxwell and Van der Waals revisited. In: Tsakalakos, T. (ed.) Proc. MRS Meeting "Phase Transformations in Solids", pp. 37–49. North-Holland, Amsterdam (1984)
3. Aifantis, E.C.: The physics of plastic deformation. Int. J. Plast. **3**, 211–247 (1987)
4. Ambrosi, D., Guillou, A.: Growth and dissipation in biological tissues. Contin. Mech. Thermodyn. **19**, 245–251 (2007)

5. Ambrosio, L., Fusco, N., Pallara, D.: Functions of Bounded Variation and Free Discontinuity Problems. Oxford University Press, Oxford (2000)
6. Ambrosio, L., Tortorelli, V.M.: Approximation of functionals depending on jumps by elliptic functionals via Γ-convergence. Commun. Pure Appl. Math. **43**, 999–1036 (1990)
7. Amor, H., Marigo, J.J., Maurini, C.: Regularized formulation of the variational brittle fracture with unilateral contact: numerical experiments. J. Mech. Phys. Solids **57**, 1209–1229 (2009)
8. Ball, J.M., James, R.D.: Fine phase mixtures as minimizers of the energy. Arch. Ration. Mech. Anal. **100**, 13–52 (1987)
9. Barenblatt, G.I.: The formation of equilibrium cracks during brittle fracture. General ideas and hypotheses. Axially-symmetric cracks. J. Appl. Math. Mech. **23**, 622–636 (1959)
10. Barenblatt, G.I.: Scaling. Cambridge University Press, Cambridge (2003)
11. Bažant, Z.P.: Instability, ductility, and size effect in strain-softening concrete. J. Eng. Mech. Div. **102**, 331–344 (1976)
12. Bažant, Z.P.: Mechanics of distributed cracking. Appl. Mech. Rev. **39**, 675–705 (1986)
13. Bažant, Z.P., Cedolin, L.: Stability of Structures. Oxford University Press, New York (1991)
14. Bažant, Z.P., Chen, E.P.: Scaling of structural failure. Appl. Mech. Rev. **50**, 593–627 (1997)
15. Bažant, Z.P., Jirásek, M.: Nonlocal integral formulations of plasticity and damage: survey of progress. J. Eng. Mech. **128**, 1119–1149 (2002)
16. Bažant, Z.P., Planas, J.: Fracture and Size Effect in Concrete and Other Quasibrittle Materials. CRC Press, Boca Raton (1998)
17. Benallal, A., Marigo, J.-J.: Bifurcation and stability issues in gradient theories with softening. Model. Simul. Mater. Sci. Eng. **15**, S283–S295 (2007)
18. Borodich, F. (ed.): Proc. IUTAM Symposium on Scaling in Solid Mechanics, Cardiff, 2007. IUTAM Bookseries, vol. 10. Springer, Berlin (2009)
19. Boscovich, R.G.: Theoria Philosophiæ Naturalis Redacta ad Unicam Legem Virium in Natura Existentium. Vienna (1758), Venice (1763), Paris (1765)
20. Bouchitté, B., Braides, A., Buttazzo, G.: Relaxation results for some free discontinuity problems. J. Reine Angew. Math. **458**, 1–18 (1995)
21. Bouchitté, B., Buttazzo, G.: Relaxation for a class of nonconvex functionals defined on measures. Ann. Inst. Henri Poincaré **10**, 345–361 (1993)
22. Bourdin, B.: Numerical implementation of the variational formulation of brittle fracture. Interfaces Free Bound. **9**, 411–430 (2007)
23. Bourdin, B., Francfort, G., Marigo, J.J.: Numerical experiments in revisited brittle fracture. J. Mech. Phys. Solids **48**, 797–826 (2000)
24. Bourdin, B., Francfort, G., Marigo, J.J.: The variational approach to fracture. J. Elast. **91**, 5–148 (2008)
25. Bourdin, B., Larsen, C., Richardson, C.L.: A time-discrete model for dynamic fracture based on crack regularization. Int. J. Fract. **168**, 133–143 (2011)
26. Braides, A.: Γ-Convergence for Beginners. Oxford University Press, Oxford (2002)
27. Braides, A., Coscia, A.: A singular perturbation approach to variational problems in fracture mechanics. Math. Models Methods Appl. Sci. **3**, 303–340 (1993)
28. Braides, A., Dal Maso, G., Garroni, A.: Variational formulation of softening phenomena in fracture mechanics: the one-dimensional case. Arch. Ration. Mech. Anal. **146**, 23–58 (1999)
29. Cahn, J.W., Hilliard, J.E.: Free energy of a uniform system. I. Interfacial energy. J. Chem. Phys. **28**, 258–267 (1958)
30. Capriz, G.: Continua with Microstructure. Springer Tracts in Natural Philosophy, vol. 35. Springer, New York (1989)
31. Carpinteri, A.: Interpretation of the Griffith instability as a bifurcation of the global equilibrium. In: Shah, S.P. (ed.) Proc. NATO Adv. Res. Workshop Application of Fracture Mechanics to Cementitious Composites, Evanston, USA, pp. 287–316. Nijhoff, Dordrecht (1985)
32. Carpinteri, A.: Cusp catastrophe interpretation of fracture instability. J. Mech. Phys. Solids **37**, 567–582 (1989)
33. Carpinteri, A.: Size effects on strength, toughness, and ductility. J. Eng. Mech. **115**, 1375–1392 (1989)
34. Carpinteri, A.: Fractal nature of material microstructure and size effects on apparent mechanical properties. Mech. Mater. **18**, 89–101 (1994)
35. Carpinteri, A., Chiaia, B., Cornetti, P.: On the mechanics of quasi-brittle materials with a fractal microstructure. Eng. Fract. Mech. **70**, 2321–2349 (2003)
36. Carpinteri, A., Cornetti, P., Puzzi, S.: Scaling laws and multiscale approach in the mechanics of heterogeneous and disordered materials. Appl. Mech. Rev. **59**, 283–305 (2006)
37. Casal, P.: La théorie du second gradient et la capillarité. C. R. Math. Acad. Sci. **274**, A1571–A1574 (1972)

38. Castellier, E., Gélébart, L., Lacour, C., Lantuéjoul, C.: Three consistent approaches of the multiple cracking process in 1D composites. Compos. Sci. Technol. **70**, 2146–2153 (2010)
39. Cattaneo, S., Rosati, G., Banthia, N.: A simple model to explain the effect of different boundary conditions in direct tensile tests. Constr. Build. Mater. **23**, 129–137 (2009)
40. Chambolle, A., Giacomini, A., Ponsiglione, M.: Crack initiation in brittle materials. Arch. Ration. Mech. Anal. **188**, 309–349 (2008)
41. Charlotte, M., Laverne, J., Marigo, J.J.: Initiation of cracks with cohesive force models: a variational approach. Eur. J. Mech. A, Solids **25**, 649–669 (2006)
42. Choksi, R., Del Piero, G., Fonseca, I., Owen, D.R.: Structured deformations as energy minimizers in models of fracture and hysteresis. Math. Mech. Solids **4**, 321–356 (1999)
43. Comi, C., Mariani, S., Negri, M., Perego, U.: A one-dimensional variational formulation for quasibrittle fracture. J. Mech. Mater. Struct. **1**, 1323–1343 (2006)
44. Coulomb, C.A.: Essai sur une application des règles *de Maximis* & *Minimis* à quelques problèmes de statique, relatifs à l'architecture. In: Mémoires de Mathématique & de Physique, présentés à l'Académie Royale des Sciences par divers Savans, vol. 7, pp. 343–382 (1773). Paris 1776
45. Dal Maso, G.: Variational problems in fracture mechanics. In: New Developments in the Calculus of Variations, Benevento, March 2005, pp. 57–67 (2006)
46. Dal Maso, G., De Simone, A., Mora, M.G., Morini, M.: Time dependent systems of generalized Young measures. Netw. Heterog. Media **2**, 1–36 (2007)
47. Dal Maso, G., Toader, R.: A model for the quasi-static growth of brittle fractures: existence and approximation results. Arch. Ration. Mech. Anal. **162**, 101–135 (2002)
48. de Borst, R., Pamin, J., Peerlings, R.H.J., Sluys, L.J.: On gradient-enhanced damage and plasticity models for failure in quasi-brittle and frictional materials. Comput. Mech. **17**, 130–141 (1995)
49. de Borst, R., Pamin, J.: Gradient plasticity in numerical simulation of concrete cracking. Eur. J. Mech. A, Solids **15**, 295–320 (1996)
50. De Giorgi, E., Ambrosio, L.: Un nuovo tipo di funzionale del calcolo delle variazioni. Atti Accad. Naz. Lincei **82**, 199–210 (1988)
51. Del Piero, G.: One-dimensional ductile-brittle transition, yielding, and structured deformations. In: Argoul, P., et al. (eds.) Proc. IUTAM Symp. Variation of Domains and Free-Boundary Problems in Solid Mechanics, pp. 203–210. Kluwer, Dordrecht (1997)
52. Del Piero, G.: Towards a unified approach to fracture, yielding, and damage. In: Inan, E., Markov, K.Z. (eds.) Proc. 9th Internat. Symposium Continuum Models and Discrete Systems, pp. 679–692. World Scientific, Singapore (1998)
53. Del Piero, G.: The energy of a one-dimensional structured deformation. Math. Mech. Solids **6**, 387–408 (2001)
54. Del Piero, G.: Bi-modal cohesive energies. In: Dal Maso, G., et al. (eds.) Variational Problems in Materials Science, Progress in Nonlinear Differential Equations and Their Applications, vol. 68, pp. 43–54. Birkhäuser, Basel (2004)
55. Del Piero, G., Lancioni, G., March, R.: A variational model for fracture mechanics: numerical experiments. J. Mech. Phys. Solids **55**, 2513–2537 (2007)
56. Del Piero, G., Lancioni, G., March, R.: A diffuse energy approach for fracture and plasticity: the one-dimensional case. J. Mech. Mater. Struct. (2013, forthcoming)
57. Del Piero, G., Owen, D.R.: Structured deformations of continua. Arch. Ration. Mech. Anal. **124**, 99–155 (1993)
58. Del Piero, G., Owen, D.R.: Integral-gradient formulae for structured deformations. Arch. Ration. Mech. Anal. **131**, 121–138 (1995)
59. Del Piero, G., Owen, D.R.: Structured Deformations. XXII Summer School of Mathematical Physics, CNR-GNFM, Ravello (1997). Quaderni dell'Istituto Nazionale di Alta Matematica (2000)
60. Del Piero, G., Owen, D.R. (eds.): Multiscale Modeling in Continuum Mechanics and Structured Deformations. CISM Courses and Lectures, vol. 447. Springer, Wien (2004)
61. Del Piero, G., Raous, M.: A unified model for adhesive interfaces with damage, viscosity, and friction. Eur. J. Mech. A, Solids **29**, 496–507 (2010)
62. Del Piero, G., Truskinovsky, L.: Macro- and micro-cracking in one-dimensional elasticity. Int. J. Solids Struct. **38**, 1135–1148 (2001)
63. Del Piero, G., Truskinovsky, L.: Elastic bars with cohesive energy. Contin. Mech. Thermodyn. **21**, 141–171 (2009)
64. Drucker, D.C.: A more fundamental approach to plastic stress-strain relations. In: Proc. 1st US Nat. Congr. Appl. Mech., pp. 487–491. ASME, New York (1951)
65. Dugdale, D.S.: Yielding of steel sheets containing slits. J. Mech. Phys. Solids **8**, 100–104 (1960)
66. Dumouchel, P.E., Marigo, J.J., Charlotte, M.: Dynamic fracture: an example of convergence towards a discontinuous quasistatic solution. Contin. Mech. Thermodyn. **20**, 1–19 (2008)

67. Ericksen, J.L.: Liquid crystals with variable degree of orientation. Arch. Ration. Mech. Anal. **113**, 97–120 (1990)
68. Fedelich, B., Ehrlacher, A.: Sur un principe de minimum concernant des matériauxà comportement indépendant du temps physique. C. R., Méc. **308**, 1391–1394 (1989)
69. Francfort, G., Marigo, J.J.: Revisiting brittle fracture as an energy minimization problem. J. Mech. Phys. Solids **46**, 1319–1342 (1998)
70. Frank, F.C.: On the theory of liquid crystals. Discuss. Faraday Soc. **25**, 19–28 (1958)
71. Freddi, F., Royer Carfagni, G.: Regularized variational theories of fracture: a unified approach. J. Mech. Phys. Solids **58**, 1154–1174 (2010)
72. Galilei, G.: Discorsi e Dimostrazioni Matematiche Intorno à Due Nuove Scienze. Elsevier, Leyden (1638)
73. Geers, M.G.D., Engelen, R.A.B., Ubachs, R.J.M.: On the numerical modelling of ductile damage with an implicit gradient-enhanced formulation. Rev. Europ. Élém. Finis **10**, 173–191 (2001)
74. Giordano, S., Mattoni, A., Colombo, L.: From elasticity theory to atomistic simulations. Rev. Comput. Chem. **27**, 1–83 (2011)
75. Goodier, J.N., Hoff, N.J. (eds.): Proc. 1st Symposium on Naval Structural Mechanics, Standford University, 1958. Pergamon, Elmsford (1960)
76. Griffith, A.A.: The phenomenon of rupture and flow in solids. Philos. Trans. R. Soc. Lond. A **221**, 163–198 (1921)
77. Gurtin, M.E., Fried, E., Anand, L.: The Mechanics and Thermodynamics of Continua. Cambridge University Press, Cambridge (2010)
78. Hackl, K., Fischer, D.F.: On the relation between the principle of maximum dissipation and inelastic evolution given by dissipation potentials. Proc. R. Soc. A, Math. Phys. Eng. Sci. **464**, 117–132 (2008)
79. Hackl, K., Hoppe, U., Kochmann, D.: Generation and evolution of inelastic microstructures—an overview. GAMM-Mitt. **35**, 91–106 (2012)
80. Hill, R.: A variational principle of maximum plastic work in classical plasticity. Q. J. Mech. Appl. Math. **1**, 18–28 (1948)
81. Hillerborg, A.: Application of the fictitious crack model to different types of materials. Int. J. Fract. **51**, 95–102 (1991)
82. Hillerborg, A., Modéer, M., Peterson, P.E.: Analysis of crack formation and crack growth in concrete by means of fracture mechanics and finite elements. Cem. Concr. Res. **6**, 773–782 (1976)
83. Hordijik, D.A.: Tensile and tensile fatigue behaviour of concrete; experiments, modelling and analyses. Heron **37**, 1–79 (1992)
84. Ippolito, M., Mattoni, A., Colombo, L., Pugno, N.: Role of lattice discreteness in brittle fracture: atomistic simulations versus analytical models. Phys. Rev. B **73**, 104111 (2006)
85. Irwin, G.R.: Analysis of stresses and strains near the end of a crack traversing a plate. J. Appl. Mech. **24**, 361–364 (1957)
86. Irwin, G.R.: Fracture mechanics. In: Goodier, J.N., Hoff, N.J. (eds.) Proc. 1st Symposium on Naval Structural Mechanics, Stanford University, 1958, pp. 557–594. Pergamon, Elmsford (1960)
87. James, R.D.: Wiggly energies. In: Batra, R.C., Beatty, M.F. (eds.) Contemporary Research in the Mechanics and Mathematics of Materials, pp. 275–286. CIMNE, Barcelona (1996)
88. Jaubert, A., Marigo, J.J.: Justification of Paris-type fatigue laws from cohesive forces model via a variational approach. Contin. Mech. Thermodyn. **18**, 23–45 (2006)
89. Jirásek, M.: Nonlocal models for damage and fracture: comparison of approaches. Int. J. Solids Struct. **35**, 4133–4145 (1998)
90. Jirásek, M., Rolshoven, S.: Localization properties of strain-softening gradient plasticity models. Part I: strain-gradient theories. Int. J. Solids Struct. **46**, 2225–2238 (2009)
91. Jirásek, M., Rolshoven, S.: Localization properties of strain-softening gradient plasticity models. Part II: theories with gradients of internal variables. Int. J. Solids Struct. **46**, 2239–2254 (2009)
92. Korteweg, D.J.: Sur la forme qui prennent les équations du mouvement des fluides si l'on tient compte des forces capillaires .... Arch. Néerl. Sci. Exactes Nat. **6**, 1–24 (1901)
93. Lancioni, G., Royer-Carfagni, G.: The variational approach to fracture mechanics. A practical application to the French Panthéon in Paris. J. Elast. **95**, 1–30 (2009)
94. Larsen, C.J.: Epsilon-stable quasi-static brittle fracture evolution. Commun. Pure Appl. Math. **63**, 630–654 (2010)
95. Larsen, C.J., Ortner, C., Süli, E.: Existence of solutions to a regularized model of dynamic fracture. Math. Models Methods Appl. Sci. **20**, 1021–1048 (2010)
96. Lasry, D., Belytschko, T.: Localization limiters in transient problems. Int. J. Solids Struct. **24**, 581–597 (1988)
97. Laverne, J., Marigo, J.J.: Approche globale, minima relatifs et critère d'amorçage en mécanique de la rupture. C. R., Méc. **332**, 313–318 (2004)

98. Lorentz, E., Andrieux, S.: A variational formulation for nonlocal damage models. Int. J. Plast. **15**, 119–138 (1999)
99. Lourenço, P.B., Almeida, J.C., Barros, J.A.: Experimental investigations of bricks under uniaxial tensile testing. Mason. Int. **18**, 11–20 (2005)
100. Lu, C.: Some notes on the study of fractals in fracture. In: Proc. 5th Australasian Congress on Applied Mechanics, ACAM 2007, Brisbane, Australia (2007)
101. Marigo, J.J.: Initiation of cracks in Griffith's theory: an argument of continuity in favor of global minimization. J. Nonlinear Sci. **20**, 831–868 (2010)
102. Marigo, J.J., Truskinovsky, L.: Initiation and propagation of fracture in the models of Griffith and Barenblatt. Contin. Mech. Thermodyn. **16**, 391–409 (2004)
103. Maxwell, J.C.: On stresses in rarified gases arising from inequalities of temperature. Philos. Trans. R. Soc. Lond. A **170**, 231–256 (1876)
104. Mielke, A.: Energetic formulation of multiplicative elasto-plasticity using dissipation distances. Contin. Mech. Thermodyn. **15**, 351–382 (2003)
105. Mielke, A.: Deriving new evolution equations for microstructures via relaxation of variational incremental problems. Comput. Methods Appl. Mech. Eng. **193**, 5095–5127 (2004)
106. Miklowitz, J.: The influence of dimensional factors on the mode of yielding and fracture in medium carbon steel-I. The geometry and size of the flat tensile bar. J. Appl. Mech. **37**, 274–287 (1948)
107. Mumford, D., Shah, J.: Optimal approximations by piecewise smooth functions and associated variational problems. Commun. Pure Appl. Math. **42**, 577–685 (1989)
108. Nemat-Nasser, S., Sum, Y., Keer, L.M.: Unstable growth of tension cracks in brittle solids: stable and unstable bifurcations, snap-through, and imperfection sensitivity. Int. J. Solids Struct. **16**, 1017–1035 (1980)
109. Orowan, E.: Energy criteria of fracture. Weld. J. **34**, 157–160 (1955)
110. Nilsson, C.: Nonlocal strain softening bar revisited. Int. J. Solids Struct. **34**, 4399–4419 (1997)
111. Owen, D.R.: Balance laws and a dissipation inequality for general constituents undergoing disarrangements and mixing. Z. Angew. Math. Mech. **88**, 365–377 (2008)
112. Pedregal, P.: Optimization, relaxation and Young measures. Bull. Am. Math. Soc. **36**, 27–58 (1999)
113. Pham, K., Marigo, J.J.: Approche variationnelle de l'endommagement: I. Les concepts fondamentaux. C. R., Méc. **338**, 191–198 (2010)
114. Pham, K., Marigo, J.J.: Approche variationnelle de l'endommagement: II. Les modélesà gradient. C. R., Méc. **338**, 199–206 (2010)
115. Pham, K., Amor, H., Marigo, J.J., Maurini, C.: Gradient damage models and their use to approximate brittle fracture. Int. J. Damage Mech. **20**, 618–652 (2011)
116. Pham, K., Marigo, J.J.: From the onset of damage to rupture: construction of responses with damage localization for a general class of gradient damage models. Contin. Mech. Thermodyn. **25**, 147–171 (2013)
117. Pugno, N.M., Ruoff, R.S.: Quantized fracture mechanics. Philos. Mag. **27**, 2829–2845 (2004)
118. Pugno, N.M.: Dynamic quantized fracture mechanics. Int. J. Fract. **140**, 159–168 (2006)
119. Rice, J.R.: The initiation and growth of shear bands. In: Palmer, A.C. (ed.) Plasticity and Soil Mechanics, pp. 263–274. Cambridge University Press, Cambridge (1973)
120. Rice, J.R.: The localization of plastic deformation. In: Koiter, W.T. (ed.) Theoretical and Applied Mechanics, vol. 1, pp. 207–220. North-Holland, Amsterdam (1976)
121. Rogula, D.: Introduction to nonlocal theory of material media. In: Rogula, D. (ed.) Nonlocal Theory of Material Media. CISM Courses and Lectures, vol. 268, pp. 125–222. Springer, Wien (1982)
122. Silling, S.A.: Reformulation of elasticity theory for discontinuities and long-range forces. J. Mech. Phys. Solids **48**, 175–209 (2000)
123. Silling, S.A., Weckner, O., Askari, E., Bobaru, F.: Crack nucleation in a peridynamic solid. Int. J. Fract. **162**, 219–227 (2010)
124. Sluckin, T.J., Dunmur, D.A., Stegemeyer, H.: Crystals That Flow. Classic Papers From the History of Liquid Crystals. Taylor & Francis, London (2004)
125. Sluys, L.J., de Borst, R.: Failure in plain and reinforced concrete—an analysis of crack width and crack spacing. Int. J. Solids Struct. **33**, 3257–3276 (1996)
126. Sorelli, L.G., Meda, A., Plizzari, G.A.: Bending and uniaxial tensile tests on concrete reinforced with hybrid steel fibers. J. Mater. Civ. Eng. **17**, 519–527 (2005)
127. Stören, S., Rice, J.R.: Localized necking in thin sheets. J. Mech. Phys. Solids **23**, 421–441 (1975)
128. Truskinovsky, L.: Fracture as a phase transition. In: Batra, R.C., Beatty, M.F. (eds.) Contemporary Research in the Mechanics and Mathematics of Materials, pp. 322–332. CIMNE, Barcelona (1996)
129. Truskinovsky, L., Zanzotto, G.: Ericksen's bar rivisited: energy wiggles. J. Mech. Phys. Solids **44**, 1371–1408 (1996)

130. Van der Waals, J.D.: Over de Continuiteit van den Gas- en Vloeistoftoestand. Ph.D. Thesis, Sijthoff, Leiden (1873). English translation: On the Continuity of the Gas and Liquid States. London (1890), and North-Holland, Amsterdam (1988)
131. Van der Waals, J.D.: The thermodynamic theory of capillarity under the hypothesis of a continuous variation of density. Verhandel. Konink. Akad. Weten. Amsterdam, Sect. 1, Vol. 1(8) (1893) (in Dutch). English translation by S. Rowlinson: J. Stat. Phys. **20**, 197–244 (1979)
132. Van Mier, J.G.M., Van Vliet, M.R.A.: Influence of microstructure of concrete on size/scale effects in tensile fracture. Eng. Fract. Mech. **70**, 2281–2306 (2003)
133. Virga, E.G.: Variational Theories for Liquid Crystals. Chapman & Hall, London (1994)
134. Volokh, K.Y.: Nonlinear elasticity for modeling fracture of isotropic brittle solids. J. Appl. Mech. **71**, 141–143 (2004)
135. Wnuk, M.P., Yavari, A.: Discrete fractal fracture mechanics. Eng. Fract. Mech. **75**, 1127–1142 (2008)
136. Yarema, S.Y.: On the contribution of G.R. Irwin to fracture mechanics. Mat. Sci. **31**, 617–624 (1995)

Index

Zeitfracht Medien GmbH
Ferdinand-Jühlke-Straße 7
99095 Erfurt, Deutschland
produktsicherheit@kolibri360.de